CRC HANDBOOK OF CHROMATOGRAPHY

Gunter Zweig and Joseph Sherma
Editors-in-Chief

GENERAL DATA AND PRINCIPLES

Editors
Gunter Zweig, Ph.D.
U.S. Environmental Protection Agency
Washington, D.C.
Joseph Sherma, Ph.D.
Lafayette College
Easton, Pennsylvania

CARBOHYDRATES

Editor
Shirley C. Churms, Ph.D.
Research Associate
C.S.I.R. Carbohydate Chemistry Research Unit
Department of Organic Chemistry
University of Cape Town, South Africa

DRUGS

Editor
Ram Gupta, Ph.D.
Department of Laboratory Medicine
St. Joseph's Hospital
Hamilton, Ontario, Canada

CATECHOLAMINES TERPENOIDS

Editor
Carmine J. Cosica, Ph.D.
Professor of Biochemistry
St. Louis University School of Medicine
St. Louis, Missouri

STEROIDS

Editor
Joseph C. Touchstone, B.S., M.S., Ph.D.
Professor
School of Medicine
University of Pennsylvania
Philadelphia, Pennsylvania

PESTICIDES

Editors
Joseph Sherma, Ph.D.
Lafayette College
Easton, Pennsylvania
Joanne M. Follweiler, Ph.D.
Lafayette College
Easton, Pennsylvania

LIPIDS AND TECHNICAL LIPID DERIVATIVES

Editor
H. K. Mangold
Executive Director and Professor
Federal Center for Lipid Research
4400 Munster, Germany

HYDROCARBONS

Editors
Willie E. May, Ph.D.
Center for Analytical Chemistry
U.S. Department of Commerce
Washington, D.C.
Walter L. Zielinski, Jr., Ph.D.
Air Program Manager
National Bureau of Standards
Washington, D.C.

INORGANICS

Editor
M. Qureshi, Ph. D.
Professor, Chemistry Section
Zakir Husain College of Engineering and Technology
Aligarh Muslim University
India

PHENOLS AND ORGANIC ACIDS

Editor
Toshihiko Hanai, Ph.D.
University of Montreal
Quebec, Canada

AMINO ACIDS AND AMINES

Editor
Dr. S. Blackburn
Leeds, England

POLYMERS

Editor
Charles G. Smith
Dow Chemical, USA
Midland, Michigan

PLANT PIGMENTS

Editor
Dr. Hand-Peter Kost
Botanisches Institut der Universitat Munchen
Munchen, West Germany

CRC Handbook of Chromatography Drugs

Volume II

Editor

Ram N. Gupta, Ph.D.
Associate Professor
Department of Pathology
McMaster University
and
Assistant Chemist
St. Joseph's Hospital
Hamilton, Ontario
Canada

Consulting Editor

Irving Sunshine, Ph.D.
Chief Toxicologist, Cuyahoga County Coroner's Office (Ohio)
Professor of Toxicology, School of Medicine
Case Western Reserve University
Cleveland, Ohio

Editors-in-Chief

Gunter Zweig, Ph.D.
School of Public Health
University of California
Berkeley, California

Joseph Sherma, Ph.D.
Professor of Chemistry
Lafayette College
Easton, Pennsylvania

CRC Press, Inc.
Boca Raton, Florida

Library of Congress Cataloging in Publication Data
Main entry under title:

Drugs.

CRC handbook of chromatography.
Bibliography: p.
Includes index.
1. Drugs--Analysis. 2. Chromatographic analysis. I. Gupta, Ram N. II. Sunshine, Irving. III. Series.
RS189.D79 615'.19015 81-10157
ISBN 0-8493-3030-0 (set) AACR2

This book represents information obtained from authentic and highly regarded sources. Reprinted material is quoted with permission, and sources are indicated. A wide variety of references are listed. Every reasonable effort has been made to give reliable data and information, but the author and the publisher cannot assume responsibility for the validity of all materials or for the consequences of their use.

Direct all inquiries to CRC Press, Inc., 2000 N.W. 24th Street, Boca Raton, Florida 33431.

International Standard Book Number 0-8493-3030-0 (Complete Set)
International Standard Book Number 0-8493-3031-9 (Volume I)
International Standard Book Number 0-8493-3032-7 (Volume II)
Library of Congress Card Number 81-10157
Printed in the United States

CRC HANDBOOK OF CHROMATOGRAPHY

Series Preface

This Handbook of Chromatography, Drugs by Ram N. Gupta, is one in a series of separate volumes devoted to a single class of chemical compounds or to compounds with a similar use pattern, like the prospective volumes on pesticides and terpenoids. When Volumes I and II of the Handbook of Chromatography were first published in 1972, the editors made an attempt to select the data so as to accomplish the coverage of most organic and inorganic compounds in a volume of about one thousand pages. However, during the ensuring ten years, the literature of chromatography, especially high-performance liquid chromatograph (HPLC), has grown to such an extent that, after an initial intent to update Volumes I and II, it was decided to publish separate volumes devoted to specific subjects. The present volume on the Chromatography of Drugs is an example of the expanded Handbook Series. In selecting Volume Editors, the Editors-in-Chief endeavored to select scientists with extensive knowledge and expertise in the chromatography of specific compounds. The Editor of this Volume, Dr. Ram N. Gupta is renowned in the field of chromatography of drugs, which is evident from the comprehensive and authoritative treatment of the subject found in this Volume. We have given each Volume Editor wide latitude in designing a format that would be most useful to the reader and do justice to the particular subject being covered. Subsequent volumes of this series will include the chromatography of pesticides, steroids, lipids and fatty acids, terpenoids, plant pigments, hydrocarbons, amino acids, inorganic compounds, polymers, and nucleic acids and associated compounds.

We invite readers to communicate with the Volume Editor for comments and corrections and to the Editors-in-Chief for suggestions for future volumes. The Editors-in-Chief want to thank Dr. Gupta for his outstanding effort and the cooperation of his associates.

Gunter Zweig, Ph.D.
Joseph Sherma, Ph.D.
Spring, 1981

PREFACE

In the last decade the most noted application of chromatography has been in the field of drug analysis. Demands for high sensitivity and selectivity in the analysis of drugs have been partly responsible for the development of sensitive and selective detectors and highly efficient columns for both gas (GC) and high-pressure liquid chromatography (HPLC).

As soon as a new drug is developed, the analytical laboratory of the pharmaceutical company proceeds to develop a sensitive analytical procedure to study the pharmacokinetics of the drug. In some cases, alternative analytical procedures are developed simultaneously in a number of centers where the drug is undergoing clinical trials. With rare exceptions, chromatographic procedures are used for the analysis of new drugs. In some instances, when the drug concentration per unit volume of the specimen is very low, immunoassays are also attempted.

In the last decade the role of the clinical chemistry laboratory has been augmented. In addition to providing analyses for different constituents both for diagnosing disease and for demonstrating symptomatic drug overdose, analyses are now being performed to monitor drug treatment. It is believed that the control of epilepsy has improved significantly with the measurement of therapeutic concentrations of antiepileptic drugs. This demand for therapeutic drug monitoring is being extended to more drug classes, e.g., antiarrhythmic drugs, antidepressants, neuroleptics, etc.

The American Society of Clinical Pathologists, the American Society of Clinical Chemists, and a number of European societies have made available voluntary quality control schemes whereby analysts can compare their performance in drug monitoring and detection with other analysts working in the field.

The spectrophotometric or colorimetric procedures used for routine quality control in the production of pharmaceuticals are being replaced by chromatographic procedures. These procedures have the capacity of detecting potentially harmful trace impurities in drugs. However, spectrophotometric or colorimetric procedures are still preferred in clinical laboratories for the rapid detection of some drugs in emergency situations, e.g., salicylates and acetaminophen.

In the last few years, there has been a phenomenal increase in publications describing the use of HPLC; however, GC is still popular and many improvements in instrumentation have been achieved. Thus, it is now quite convenient to use nitrogen-selective detectors in which the salt bead is heated electrically. With the introduction of flexible break-resistant capillary columns, GC has offered a new potential for ultratrace analysis in complex matrices. Another factor in favor of GC is that the mobile phase does not present a disposal problem.

In North America, quantitative thin-layer chromatography (TLC) is not as popular as it is in Europe. One advantage of TLC is the ability to separate a number of samples simultaneously, and the separated spots can be quantitated relatively rapidly by *in situ* densitometry. The spots can be made colored or fluorescent by spraying or dipping the plate in suitable reagents. Postcolumn reactions in HPLC to increase the detection sensitivity are still not very popular. TLC remains the method of choice all over the world for the qualitative detection of drugs of abuse. There have been advances in the manufacture of precoated TLC plates. High-performance TLC plates and chemically bonded reverse-phase TLC plates are now commercially available.

The purpose of this handbook is to provide a reference source of different chromatographic techniques available for the analysis of drugs.

I am grateful to Mrs. Elaine Moore for her skillful assistance in preparing the manuscript. Mrs. Diane Lewis helped to organize the filing system of reprints. Ms. Pamela Woodcock of CRC Press provided all the required editorial assistance.

Ram N. Gupta

THE EDITORS-IN-CHIEF

Gunter Zweig, Ph.D., received his undergraduate and graduate training at the University of Maryland, where he was awarded the Ph.D. in biochemistry in 1952. For two years after his graduation, Dr. Zweig was affiliated with the late R. J. Block, pioneer in paper chromatography of amino acids. Zweig, Block and Le Strange wrote one of the first books on paper chromatography which was published in 1952 by Academic Press and went into three editions, the last one authored by Gunter Zweig and Dr. Joe Sherma, the co-Editor-in-Chief of this Handbook. Paper Chromatography (1952) was also translated into Russian.

From 1953 till 1957, Dr. Zweig was research biochemist at the C. F. Kettering Foundation, Antioch College, Yellow Springs, Ohio, where he pursued research on the path of carbon and sulfur in plants using the then newly developed techniques of autoradiography and paper chromatography. From 1957 till 1965, Dr. Zweig served as lecturer and chemist, University of California, Davis and worked on analytical methods for pesticide residues, mainly by chromatographic techniques. In 1965, Dr. Zweig became Director of Life Sciences, Syracuse University Research Corporation (research on environmental pollution), and in 1973 he became Chief, Environmental Fate Branch, Environmental Protection Agency in Washington, D.C. In 1980, he was appointed Senior Science Advisor in the same agency. During his government career, Dr. Zweig continued his scientific writing and editing. Among his works are (many in collaboration with Dr. Sherma) the now 11-volume series on Analytical Methods for Pesticides and Plant Growth Regulators (Academic Press); the Pesticide Chemistry series for CRC Press; co-editor of Journal of Toxicology and Environmental Health; co-author of basic review on paper and thin-layer chromatography for Analytical Chemistry from 1968-1980; co-author of applied chromatography review on pesticide analysis for Analytical Chemistry, beginning in 1981. Among the scientific honors awarded to Dr. Zweig during his distinguished career are the Wiley Award in 1977, Rothschild Fellowship to the Weizmann Institute in 1963/64; the Bronze Medal by the EPA in 1980. Dr. Zweig has authored or co-authored over 75 scientific papers on diverse subjects in chromatography and biochemistry, besides being the holder of three U.S. patents. At the present time (1980/82), Dr. Zweig is Visiting Scholar in the School of Public Health, University of California, Berkeley, where he is doing research on farmworker safety as related to pesticide exposure.

Joseph Sherma, Ph.D., received a B.S. in Chemistry from Upsala College, East Orange, N.J., in 1955 and a Ph.D. in Analytical Chemistry from Rutgers University in 1958. His thesis research in ion exchange chromatography was under the direction of the late William Rieman III. Dr. Sherma joined the faculty of Lafayette College in September, 1958, and is presently full professor there in charge of two courses in analytical chemistry. At Lafayette he has continued research in chromatography and has additionally worked a total of 12 summers in the field with Harold Strain at the Argonne Nationa Laboratory, James Fritz at Iowa State University, Gunter Zweig at Syracuse University Research Corporation, Joseph Touchstone at the Hospital of the University of Pennsylvania, Brian Bidlingmeyer at Waters Associates, and Thomas Beesley at Whatman, Inc. Dr. Sherma and Dr. Zweig (who is now with U.S. EPA) co-authored Volumes I and II of the *CRC Handbook of Chromatography,* a book on paper chromatography, and 6 volumes of the series *Analytical Methods for Pesticides and Plant Growth Regulators.* Other books in the pesticide series and further volumes of the *CRC Handbook of Chromatography* are being edited with Dr. Zweig, and Dr. Sherma will co-author the Handbook on Pesticide Chromatography. A book on quantitative TLC (Wiley-Interscience) was edited jointly with Dr. Touchstone. Dr. Sherma has been co-author of seven biennial reviews of liquid chromatography (1968-1980) and the 1981 review of pesticide analysis for the journal *Analytical Chemistry.* Dr. Sherma has authored major invited chapters and review papers on chromatography and pesticides in *Chromatographic Reviews* (analysis of fungicides), *Advances in Chromatography* (analysis of non-pesticide pollutants), Heftmann's *Chromatography* (chromatography of pesticides), Race's *Laboratory Medicine* (chromatography in clinical analysis), *Food Analysis: Principles and Techniques* (TLC for food analysis), *Treatise on Analytical Chemistry* (paper and thin layer chromatography), and *CRC Critical Reviews in Analytical Chemistry* (pesticide residue analysis). A general book on thin layer chromatography co-authored by Dr. Sherma is now in press at Marcel Dekker. Dr. Sherma spent six months in 1972 on sabbatical leave at the EPA Perrine Primate Laboratory, Perrine, Florida, with Dr. T. M. Shafik, and two additional summers (1975, 1976) at the USDA in Beltsville, Maryland, with Melvin Getz doing research on pesticide residue analysis methods development. He spent three months in 1979 on sabbatical leave with Dr. Touchstone developing clinical analytical methods. A total of more than 200 papers, books, book chapters, and oral presentations concerned with column, paper, and thin layer chromatography of metal ions, plant pigments, and other organic and biological compounds; the chromatographic analysis of pesticides; and the history of chromatography have been authored by Dr. Sherma, many in collaboration with various co-workers and students. His major research area at Lafayette is currently quantitative TLC (densitometry), applied mainly to clinical analysis and pesticide residue determinations. Dr. Sherma has written an analytical quality control manual for pesticide analysis under contract with the U.S. EPA and has revised this and the EPA Pesticide Analytical Methods Manual under a four-year contract (EPA) jointly with Dr. M. Beroza of the AOAC. Dr. Sherma has also written an instrumental analysis quality assurance manual and other analytical reports for the U.S. Consumer Product Safety Commission, and is currently preparing a manual on the analysis of food additives for the U.S. FDA, both of these projects also in collaboration with Dr. Beroza of the AOAC. Dr. Sherma taught the first, prototype short course on pesticide analysis, with Henry Enos of the EPA, for the Center for Professional Advancement. He is editor of the Kontes TLC quarterly newsletter and also teaches short courses on TLC for Kontes and the Center for Professional Advancement. He is a consultant for several industrial companies and federal agencies on chemical analysis and chromatography and regularly referees papers for analytical journals and research proposals for government agencies. Dr. Sherma has received two awards for superior teaching at Lafayette College and the 1979 Distinguished Alumnus Award from Upsala College for outstanding achievements as an educator, researcher, author, and editor. He is a member of the ACS, Sigma Xi, Phi Lambda Upsilon, SAS, and AIC.

THE EDITOR

Ram N. Gupta, Ph.D., is Assistant Clinical Chemist at the St. Joseph's Hospital in Hamilton and Associate Professor in the Department of Pathology at McMaster University, in Hamilton.

Dr. Gupta received his M.Sc. degree in 1962 and Ph.D. degree in 1963 in Organic Chemistry from McMaster University. He continued working in the Chemistry Department of McMaster University as a research associate until 1971 when he moved to the Department of Pathology at the same university.

Dr. Gupta has been elected as a fellow of the Chemical Institute of Canada. He is a member of the American Chemical Society, American Association of Clinical Chemists, Canadian Society of Clinical Chemists and the Association of Clinical Biochemists (U.K.). He is the author of more than 40 scientific publications.

His present research interests are the development of chromatographic procedures for the assay of drugs and other biochemicals in biological fluids.

THE CONSULTING EDITOR

Dr. Irving Sunshine is Chief Toxicologist at the Cuyahoga County (Cleveland), Ohio Coroner's Office; Professor of Toxicology in the Department of Pathology and Professor of Clinical Pharmacology in the Department of Medicine at the School of Medicine, Case Western Reserve University; Chief Toxicologist for the University Hospitals in Cleveland, Ohio; Director of the Cleveland Poison Information Center; and Editor-In-Chief for Biosciences for CRC Press, Inc. He is a Diplomate of both the American Board of Clinical Chemistry and The American Board of Forensic Toxicology and is on the Board of Directors of both these organizations.

Born in New York City, he obtained all his formal education in various Colleges of New York University, earning the B.Sc., M.A., and Ph.D. degrees. While earning his Ph.D., he taught chemistry in various colleges in the New York area and during the war, he worked during the "grave yard" shift on a pilot plant for the separation of uranium isotopes as a part of the "The Manhattan Project". His development in toxicology was encouraged by two memorable mentors, Dr. Alexander O. Gettler and Dr. Bernard Brodie.

To my teacher

Professor Ian D. Spenser

TABLE OF CONTENTS

ORGANIZATION OF TABLES AND EXPLANATION OF ABBREVIATIONS

The drugs have been arranged alphabetically according to their generic names. Some of the commonly used synonyms and proprietary names for them are listed in the Appendix and included in the Index. Drugs for which a suitable chromatographic procedure could not be found in the current literature are not included in the handbook. Steroidal drugs will be included in another volume of this series of handbooks.

Gas Chromatography (GC)

Specimen — B = whole blood, plasma, or serum; U = urine; D = dosage form (pharmaceutical preparation, plant material, etc.); NA = not available; CSF = cerebrospinal fluid. Other kinds of specimens have been described without using abbreviations.

S — Sensitivity of the procedure. We have categorized a procedure as 1. if the authors have used that procedure for a single-dose pharmacokinetic study; as 2. if the procedure has been used to measure concentration after multiple dosage. In many instances, a procedure categorized as 2. may be adequate for pharmacokinetic study. Category 3. has been assigned to those procedures which have been used to analyze dosage forms or to obtain a semiquantitative or a qualitative result.

Column — Columns are made of glass unless noted otherwise. Length is given in meters and inner diameter in millimeters.

Packing — In a few cases, the chemical names of liquid phases have been changed to commonly used abbreviations; support size if given in other units has been changed to mesh sizes.

Oven temp. — Oven temperature is given for isothermal operation only. It is merely indicated that the procedure uses temperature programming.

Gas — Gas flow if given in units other than mℓ/min has been indicated by a footnote.

Det. — Detector. FID= Flame Ionization Detector; NPD = Nitrogen Phosphorus Detector, Alkali Flame Ionization Detector, Thermionic Sensitive Detector, or Nitrogen Specific Detector; ECD = Electron Capture Detector; MS-EI = Electron-Impact Mass Spectrometer, and MS-CI = Chemical Ionization Mass Spectrometer. Any other detector used has been indicated without using abbreviations.

RT (min) — Retention time in minutes. This column gives the retention time of the title drug as it appears in the chromatogram of the procedure. It may be the retention time of the parent drug or its derivative if formed during the treatment of derivatization process. A dash (—) indicates that the title drug is not analyzed in the procedure under review, whereas NA indicates that the retention time is not available.

Internal Standard — The names of the compounds used as internal standards are given in full. Any abbreviation used is explained by a footnote. A dash (—) indicates that no internal standard was used in the procedure. The retention time in minutes is given in parentheses as it appears in the chromatogram. It may be of the parent compound or its derivative if formed during the derivatization process.

Deriv. — Derivative. This column indicates that the specimen or its extract at some stage has been subjected to derivatization. A footnote indicates the derivatizing agent when a number of alternative reagents are available to prepare a particular derivative.

Other Compounds (RT) — Metabolites of the parent drug or other similar or unrelated drugs, when analyzed simultaneously with the title drug are listed in this column. Their retention times in minutes are given in parentheses.

Ref. — Reference.

High-Pressure Liquid Chromatography (HPLC)* (See under GC for the explanation of common columns)

Column — Columns are made of steel unless noted otherwise. Length is in centimeters and inner diameter in millimeters. Temperature other than ambient is indicated by a footnote.

Packing — Packing is described by the trade names as used by the authors.

Elution — Unless noted otherwise the procedure uses isocratic elution. In this column, the eluting solvent is given a number and the corresponding solvent is described at the end of the table.

Flow Rate — Flow rate given in other units has been changed to mℓ/min; a footnote indicates that only the pump pressure is given.

Det. — Detection. Wavelength (nm) for ultraviolet (UV) or visible absorption is given. Fl = fluorescence; λ_{ex} = excitation wavelength, λ_{fl} = emission wavelength. Other modes of detection are indicated without the use of abbreviations.

Thin-Layer Chromatography (TLC) (See under GC and HPLC for the explanation of common columns)

Plate — Unless otherwise noted, plates are made of glass. "Laboratory" indicates that the plates have been coated by the authors in their laboratory. A few cases of paper chromatography have also been included under TLC. The manufacturer of paper and the kind of paper used are indicated under plate.

Solvent — Developing solvent is given a number, and the corresponding solvent is described at the end of the table.

Post-Separation Treatment — Sp: The plate is sprayed with the described reagent. D: The plate is dipped in the described reagent. E: The plate is exposed to the vapors of the described reagent.

Det. — Qualitative detection is indicated as Visual. Wavelength (nm) for short or long wave UV lamp is given when fluorescence or quenching of fluorescence is observed under a UV light. When the plate is scanned with a densitometer for quantitative analysis, the mode of scanning is indicated as reflectance, transmission or reflectance/transmission for simultaneous mode. Wavelength (nm) for scanning mode is given and for fluorescence scanning both excitation wavelength (λex) and emission wavelength (λfl) are given.

* Also known as High Performance Liquid Chromatography.

IBUPROFEN

Gas Chromatography

Specimen (mℓ)	S	Column m × mm	Packing (mesh)	Oven temp (°C)	Gas (mℓ/min)	Det.	RT (min)	Internal standard (RT)	Deriv.	Other compounds (RT)	Ref.
B (1)	2	1.8 × 2	5% FFAP Gas Chrom® W (80/100)	220	Nitrogen (35)	FID	4.3	3-Methyl-3-phenylbutyric acid (2.9)	—	—	1
B (0.1)	1	1.8 × 3	10% 3-Cyanopropyl-silicone Gas Chrom® Q (100/120)	190	Argon 95 + Methane (55)	ECD	9.1	4-Isobutylphenyl acetic acid (13.3)	Pentafluorobenzyl	—	2
B (2)	2	1.8 × 4	10% SP-216-PS Supelcoport (100/120)	200	Nitrogen (80)	FID	8.9	Myristic acid (3.6)	—	—	3

High-Pressure Liquid Chromatography

Specimen (mℓ)	S	Column cm × mm	Packing (μm)	Elution	Flow rate (mℓ/min)	Det. (nm)	RT (min)	Internal standard (RT)	Other compounds (RT)	Ref.
B (1)	2	25 × 4.6	LiChrosorb® RP-18 (10)	E-1	2	UV (220)	6.5	Cinnamic acid (2.7)	—	4

Note: E-1 = 0.01 *M* Orthophosphoric acid + methanol.
3 : 7

IBUPROFEN (continued)

REFERENCES

1. Hoffman, D. J., *J. Pharm. Sci.*, 66, 749, 1977.
2. Kaiser, D. G. and Martin, R. S., *J. Pharm. Sci.*, 67, 627, 1978.
3. Hackett, L. P. and Dusci, L. J., *Clin. Chim. Acta*, 87, 301, 1978.
4. Pitre, D. and Grandi, M., *J. Chromatogr.*, 170, 278, 1979.

IBUPROXAM

Gas Chromatography

Specimen (mℓ)	S	Column m × mm	Packing (mesh)	Oven temp (°C)	Gas (mℓ/min)	Det.	RT (min)	Internal standard (RT)	Deriv.	Other compounds (RT)	Ref.
B (5)	2	2 × 4	6% Dexsil® 300 GC Chromosorb® W (NA)	200	Nitrogen (60)	FID	NA	—	Methylsilyl	Ibuprofen (NA)	1[a]

[a] Analysis has also been carried out by thin-layer chromatography.

REFERENCE

1. Orzalesi, G., Selleri, R., Caldini, O., Volpato, I., Innocenti, F., Colome, J., Sacristan, A., Varez, G., and Pisaturo, G., *Arzneim. Forsch.*, 27, 1012, 1977.

IMIPRAMINE

Gas Chromatography

Specimen (mℓ)	S	Column m × mm	Packing (mesh)	Oven temp (°C)	Gas (mℓ/min)	Det.	RT (min)	Internal standard (RT)	Deriv.	Other compounds (RT)	Ref.
B (0.5—2)	2	1.5 × 1.5	3% OV-17 Supelcoport (80/100)	245	Nitrogen (37.5)	NPD	1.7	N-7084[a] (3.8)	Trifluroacetyl	Desipramine (3.3)	1
B (1—3)	2	1.8 × 2	5% OV-17 Supelcoport (100/120)	240	Helium (22)	NPD	4.8	Amitriptyline (4.3)	—	Desipramine (5.7) 2-Hydroxyimipramine (12.9) 2-Hydroxydesipramine (19)	2
B (1)	2	1 × 2	3% OV-1 Gas Chrom® Q (100/120)	240	Helium (20)	MS-EI	1	Promazine (1.6)	Acetyl	Desipramine (2.8)	3
B (3)	2	1.8 × 2	3% OV-17 Gas Chrom® Q (100/120)	Temp. Progr.	Helium (17)	NPD	4.6	Promazine (6.1)	Trifluoroacetyl	Desipramine (7)	4
B (2)	2	1.2 × 2	3% OV-17 Gas Chrom® Q (80/100)	Temp. Progr.	Helium (35)	NPD	4.1	Butriptyline (3.5)	—	Desipramine (4.6)	5

High-Pressure Liquid Chromatography

Specimen (mℓ)	S	Column cm × mm	Packing (μm)	Elution	Flow rate (mℓ/min)	Det. (nm)	RT (min)	Internal standard (RT)	Other compounds (RT)	Ref.
B (2)	2	30 × 3.9	μ-Bondapak® C_{18} (10)	E-1	1.5	UV (254)	11.2	β-Naphthylamine (4.5)	Desipramine (9) Doxepin (8) Amitriptyline (13)	6

B (2)	2	30 × 4.6	LiChrosorb® Si 60 (5)	E-2	1.3	UV 254)	13	Promazine (19)	Amitriptyline (10)	7
B (1)		25 × 4.6	Silica (5)	E-3	1.8	Fl (Ex: 240, Fl: 370)	7	Desmethylclomipramine (15.1)	Desipramine (9.2) 2-Hydroxyimipramine (8.1) 2-Hydroxydesipramine (20.3)	8
B (2)	1	30 × 3.9	μ-Bondapak®[b] Phenyl (10)	E-4	2	Fl (Ex: 252, Fl: 360)	11	Trimipramine (14)	Desipramine (10.5)	10

Thin-Layer Chromatography

Specimen (mℓ)	S	Plate (manufacturer)	Layer (mm)	Solvent	Post-separation treatment	Det. (nm)	R_f	Internal standard (R_f)	Other compounds (R_f)	Ref.
B (5)	2	20 × 20 cm (Author) (Laboratory)	Silica gel G (0.25)	S-1	Ex: Nitrous gases	Reflectance/ transmission (405)	0.35	—	Desipramine (0.12)	9

Note: E-1 = Methanol + acetonitrile + 0.1 *M* phosphate buffer, pH 7.6.
41 : 15 : 44

E-2 = Ethyl acetate (dry) + methylamine.
100 : .05

E-3 = Methanol + acetonitrile + ammonium hydroxide.
1 : 5 : 4 mℓ/ℓ

E-4 = Acetonitrile + 0.015% phosphoric acid.
79 : 21

S-1 = Chloroform + diethyl ether + methanol.
85 : 15 : 20

[a] 5-(3-Pyrrolidin-1-yl-propylidene)-10,11 dihydro-5-H-dibenzo(ad)cycloheptane.
[b] Column Temperature = 50°C.

IMIPRAMINE (continued)

REFERENCES

1. Reite, S. F., *Medd. Nor. Farm. Selsk.*, 37, 76, 1975.
2. Cooper, T. B., Allen, D., and Simpson, G. M., *Psychopharmacol. Commun.*, 1, 445, 1975.
3. Belvedere, G., Bruti, L., Frigerio, A., and Pantarotto, C., *J. Chromatogr.*, 111, 313, 1975.
4. Bailey, D. N. and Jatlow, P. I., *Clin. Chem.*, 22, 1697, 1976.
5. Bertrand, M., Dupuis, C., Gagnon, M.-A., and Dugal, R., *Clin. Biochem.*, 11, 117, 1978.
6. Proelss, H. F., Lohman, H. J., and Miles, D. G., *Clin. Chem.*, 24, 1948, 1978.
7. van den Berg, J. H. M., de Ruwe, H. J. J. M., and Deelder, R. S., *J Chromatogr.*, 138, 431, 1977.
8. Sutfin, T. A. and Jusko, W. J., *J. Pharm. Sci.*, 68, 703, 1979.
9. Nagy, A. and Treiber, L., *J. Pharm. Pharmacol.*, 25, 599, 1973.
10. Reece, P. A., Zacest, R., and Barrow, C. G., *J. Chromatogr.*, 163, 310, 1979.

INDICINE

Gas Chromatography

Specimen (mℓ)	S	Column m × mm	Packing (mesh)	Oven temp (°C)	Gas (mℓ/min)	Det.	RT (min)	Internal standard (RT)	Deriv.	Other compounds (RT)	Ref.
B, U (1)	2	1.2 × 2	3% OV-101 Supelcoport (80/100)	185	Argon 95 + Methane 5 (30)	ECD	4.2	Heliotrine + Heliotrine-*N*-oxide[a] (7)	Pentafluoropropionyl	Indicine-*N*-oxide[a] (4.2)	1
U (1)	2	1.2 × 2	3% OV-17 Supelcoport (80/100)	Temp. Progr.	NA	MS-EI	NA[b]	Heliotrine-*N*-oxide (NA)[b] + Echinatine (NA)[b]	Trimethylsilyl	Indicine-*N*-oxide[b] (NA)	2

[a] *N*-oxides are converted to corresponding tertiary bases by reduction with zinc dust and acetic acid.
[b] Retention data is given in methylene units.

REFERENCE

1. **Ames, M. M. and Powis, G.**, *J. Chromatogr.*, 166, 519, 1978.
2. **Evans, J. V., Peng, A., and Nielsen, C. J.**, *Biomed. Mass Spectrom.*, 6, 38, 1979.

INDOMETHACIN

Gas Chromatography

Specimen (mℓ)	S	Column m × mm	Packing (mesh)	Oven temp (°C)	Gas (mℓ/min)	Det.	RT (min)	Internal standard (RT)	Deriv.	Other compounds (RT)	Ref.
B (1)	2	1 × 2.5	2% OV-1 Chromosorb® W (100/120)	190	Nitrogen (60)	ECD	9	—	Ethyl[a]	—	1
B, U (0.5)	2	1.5 × 4	2% SE-52 Chromosorb® W (100/120)	245	Nitrogen (60)	ECD	10	—	Ethyl[a]	—	2
B[b] (0.5)	2	1.2 × 3	3% SE-52 Supelcoport (100/120)	260	Argon 90 + Methane 10 (80)	ECD	3.5	4-Flurobenzoyl analog (2)	Methyl[c]	—	3
B (NA)	2	1.8 × 2	3% OV-1 Gas Chrom® Q (80/100)	240	Nitrogen (30)	FID	10.6	Methylindomethacin (7.3)	Propyl[d]	—	4
B, U (1)	2	1.5 × 4	2% Dexsil® 300 Chromosorb® W (100/120)	305	Nitrogen (45)	ECD	4.2	5-Fluroindomethacin (2.4)	Pentafluorobenzyl	—	5

High-Pressure Liquid Chromatography

Specimen (mℓ)	S	Column cm × mm	Packing (μm)	Elution	Flow rate (mℓ/min)	Det. (nm)	RT (min)	Internal standard (RT)	Other compounds (RT)	Ref.
B (1)	2	61 × 2.1	μ-Bondapak® C_{18} Corasil (37—50)	E-1	1	UV (254)	5.7	Flufenazinic acid (3.3)	—	6
B (0.1)	2	30 × 4	μ-Bondapak® C_{18} (10)	E-2	3	UV (200)	2.3	2,5-Diphenyloxazole (4)	—	7

Thin-Layer Chromatography

Specimen (mℓ)	S	Plate (manufacturer)	Layer (mm)	Solvent	Post-separation treatment	Det. (nm)	R_f	Internal standard (R_f)	Other compounds (R_f)	Ref.
B, U (1)	2	20 × 20 cm (Merck)	Silica gel 60 (0.25)	S-1[f]	—	UV-Reflectance/ transmission (330)	0.37	—	—	8

Note: E-1 = Acetonitrile + 0.01 *M* acetic acid.
40 : 60

E-2 = 10 m*M* Potassium phosphate buffer + acetonitrile.
40 : 60

S-1 = Chloroform + methanol.
30 : 6

[a] With diazoethane.
[b] And aqueous humor.
[c] With diazomethane.
[d] With propyl iodide in the presence of tetrabutylammonium hydrogen sulfate, hexacosane, and solid sodium bicarbonate.
[e] Column temperature = 40°C.
[f] Development was carried out in dark.

REFERENCES

1. **Ferry, D. G., Ferry, D. M., Moller, P. W., and McQueen, E. G.,** *J. Chromatogr.*, 89, 110, 1974.
2. **Helleberg, L.,** *J. Chromatogr.*, 117, 167, 1976.
3. **Plazonnet, B. and Vandenheuvel, W. J. A.,** *J. Chromatogr.*, 142, 587, 1977.
4. **Arbin, A.,** *J. Chromatogr.*, 144, 85, 1977.
5. **Sibeon, R. G., Baty, J. D., Baker, N., Chan, K., and Orme, L. E.,** *J. Chromatogr.*, 153, 189, 1978.
6. **Skellern, G. G. and Salole, E. G.,** *J. Chromatogr.*, 114, 483, 1975.

INDOMETHACIN (continued)

7. Soldin, S. J. and Gero, T., *Clin. Chem.*, 25, 589, 1979.
8. Søndergaard, I. and Steiness, E., *J. Chromatogr.*, 162, 485, 1979.

INDOPROFEN

Gas Chromatography

Specimen (mℓ)	S	Column m × mm	Packing (mesh)	Oven temp (°C)	Gas (mℓ/min)	Det.	RT (min)	Internal standard (RT)	Deriv.	Other compounds (RT)	Ref.
B, U (2)	1	2 × 4	1% SE-54 Chromosorb® W (60/80)	240	Nitrogen (60)	FID	3.5	[4-(1-OXO-2-isoindolyl)-phenyl] pentanoic acid[a] (5)	Trifluoroethyl ester	—	1
B (2)	1	0.6 × 4	3% OV-1 Chromosorb® W (100/120)	235	Nitrogen (30)	FID	2.5	[4-(1-OSO-2-isoindolyl)-phenyl] pentanoic acid (4.5)	Methyl[b]	—	2

High-Pressure Liquid Chromatography

Specimen (mℓ)	S	Column cm × mm	Packing (μm)	Elution	Flow rate (mℓ/min)	Det. (nm)	RT (min)	Internal standard (RT)	Other compounds (RT)	Ref.
B, U (NA)	1	30 × 4	μ-Bondapak® C_{18} (10)	E-1	1.0	UV (280)	5.5	[4-(1-OXO-2-isoindolyl)-phenyl] pentanoic acid (10.5)	—	3

Note: E-1 = Acetonitrile + 0.175 *M* acetic acid.
40 : 60

[a] A second quantitation standard — trifluoroethyl ester of [4-(1-OXO-2-isoindolyl)phenyl] hexanoic acid — is added just prior to chromatography (retention time = 6.5).

[b] With diazomethane.

INDOPROFEN (continued)

REFERENCES

1. Tosolini, G. P., Forgione, A., Moro, E., and Mandelli, V., *J. Chromatogr.*, 92, 61, 1974.
2. Smith, R. V., Humphrey, D. W, and Escalona—Castillo, H., *J. Pharm. Sci.*, 66, 132, 1977.
3. Lakings, D. B., Haggerty, W. J., Rehagen, G., and Barth, H., *J. Pharm. Sci.*, 67, 831, 1978.

IODOCHLORHYDROXYQUIN

Gas Chromatography

Specimen (mℓ)	S	Column m × mm	Packing (mesh)	Oven temp (°C)	Gas (mℓ/min)	Det.	RT (min)	Internal standard (RT)	Deriv.	Other compounds (RT)	Ref.
B (1)	2	1.1 × 3	2% XF 1105 Chromosorb® W (60/80)	175	Nitrogen (85)	ECD	7	5,7-Dichloro-8-quinolinol (3.5)	Acetyl	—	1
B (0.1—0.5)	2	1.5 × 2	3% JXR Gas Chrom® Q (NA)	185	Nitrogen (30)	ECD	2	5,7-Dichloro-8-quinolinol (0.9)	Methyl[a]	—	2
B (0.1—0.5)	2	1.5 × 2	3% OV-17 Gas Chrom® Z (80/100)	215	Nitrogen (30)	ECD	4	Chloroquinaldol (NA)	Methyl[b]	—	3

Thin-Layer Chromatography

Specimen (mℓ)	S	Plate (manufacturer)	Layer (mm)	Solvent	Post-separation treatment	Det. (nm)	R_f	Internal standard (R_f)	Other compounds (R_f)	Ref.
B (1.5)	1	NA (Merck)	Silica gel 60 F_{254} (0.25)	S-1	—	i) Visual (254) ii) uv[d] (267)	0.58	—	—	4
D	3	20 × 25 cm (Laboratory)	Silica gel 60 HR-F_{254} (0.25)	S-2[c]	—	i) Visual (UV-254) ii) UV[d] (269)	0.40	—	5,7-Diodo-8-hydroxyquinoline (0.35) 5-Chloro-8-hydroxyquinoline (0.76) 8-Hydroxyquinoline (0.87)	5

IODOCHLORHYDROXYQUIN (continued)

Note: S-1 = 1-Butanol + acetone + diethylamine + water.
30 : 20 : 4 : 30

S-2 = Triethylamine + dioxane + methyl ethyl ketone.
80 : 15 : 5

[a] Extractive alkylation with methyl iodide and tetrahexylammonium hydrogen sulfate.
[b] Extractive alkylation with methyl iodide and tetrabutylammonium hydroxide.
[c] The plate is developed three times in the same solvent.
[d] The separated spots are scraped, eluted, and determined spetrophotometrically.

REFERENCES

1. **Tamura, Z., Yoshioka, M., Imanari, T., Fukaya, J., Kusaka, J., and Samejima, K.,** *Clin. Chim. Acta,* 47, 13, 1973.
2. **Degen, P. H., Schneider, W., Vuillard, P., Geiger, U. P., and Riess, W.,** *J. Chromatogr.,* 117, 407, 1976.
3. **Hartvig, P. and Fagerlund, C.,** *J. Chromatogr.,* 140, 170, 1977.
4. **Soldato, P. D.,** *J. Pharm. Sci.,* 66, 1334, 1977.
5. **Valle, O. R., Jimenez, D., Lopez, S. G., and Schroeder, I.,** *J. Chromatogr. Sci.,* 16, 162, 1978.

IPROCLOZIDE

Gas Chromatography

Specimen (mℓ)	S	Column m × mm	Packing (mesh)	Oven temp (°C)	Gas (mℓ/min)	Det.	RT (min)	Internal standard (RT)	Deriv.	Other compounds (RT)	Ref.
U (200)	2	2.5 × 2	5% OV-225 Gas Chrom® Q (80/100)	190	Nitrogen[a]	FID	28	2-Butyl analog (40)	—	—	1

Thin-Layer Chromatography

Specimen (mℓ)	S	Plate (manufacturer)	Layer (mm)	Solvent	Post-separation treatment	Det. (nm)	R_f	Internal standard (R_f)	Other compounds (R_f)	Ref.
D	3	20 × 20 cm (Laboratory)	Cellulose HN-300 + Silica gel HF-254 (0.25)	S-1[b]	Folin-Ciocalteu reagent[c]	Visual	0.81	—	Isoniazid (0.18) Iproniazid (0.60) Isocarboxazid (0.81) Nialamide (0.45)	2

Note: S-1 = Chloroform + methanol + 13 *N* ammonia.
90 : 10 : 1

[a] Linear velocity = 14.3 cm/sec.
[b] Seven different solvent systems have been described.
[c] Thirteen spray reagents have been evaluated.

REFERENCES

1. de Sagher, R. M., de Leenheer, A. P., and Claeys, A. E., *Anal. Chem.*, 47, 1144, 1975.
2. de Sagher, R. M., de Leenheer, A. P., and Claeys, A. E., *J. Chromatogr.*, 106, 357, 1975.

IPRONIAZID

Gas Chromatography

Specimen (mℓ)	S	Column m × mm	Packing (mesh)	Oven temp (°C)	Gas (mℓ/min)	Det.	RT (min)	Internal standard (RT)	Deriv.	Other compounds (RT)	Ref.
U (200)	2	2.5 × 2	5% XE-60 Gas Chrom® Q (80/100)	180	Nitrogen[a]	FID	6.5	2-Butyl analog (9.5)	—	—	1

[a] Linear velocity = 14.3 cm/sec.

REFERENCE

1. de Sagher, R. M., de Leenheer, A. P., and Claeys, A. E., *J. Pharm. Sci.*, 65, 878, 1976.

ISAMOXOLE

Gas Chromatography

Specimen (mℓ)	S	Column m × mm	Packing (mesh)	Oven temp (°C)	Gas (mℓ/min)	Det.	RT (min)	Internal standard (RT)	Deriv.	Other compounds (RT)	Ref.
B (0.5—1)	1	2 × NA	3% OV-1 Gas Chrom® Q (80/100)	165	Helium (30)	MS-EI	2.5	[$^2H_9$] Butyl analog (2.5)	—	—	1

REFERENCE

1. Chatfield, D. H. and Woodage, T. J., *Biomed. Mass Spectrom.*, 5, 466, 1978.

ISOFLUROPHATE

Gas Chromatography

Specimen (mℓ)	S	Column m × mm	Packing (mesh)	Oven temp (°C)	Gas (mℓ/min)	Det.	RT (min)	Internal standard (RT)	Deriv.	Other compounds (RT)	Ref.
D	3	1.5 × 4	4% OF-1 Gas Chrom® Q (80/100)	105	NA (40)	FID	—	—	Methyl[a]	Diisopropylphosphate[b] (3.8)	1

[a] With diazomethane.
[b] Decomposition product of isoflurophate.

REFERENCE

1. Ryan, J. A., McGaughran, W. R., Lindemann, C. J., and Zacchei, A. G., *J. Pharm. Sci.*, 68, 1194, 1979.

ISONIAZID

Gas Chromatography

Specimen (mℓ)	S	Column m × mm	Packing (mesh)	Oven temp (°C)	Gas (mℓ/min)	Det.	RT (min)	Internal standard (RT)	Deriv.	Other compounds (RT)	Ref.
U (25)	2	2 × 1.8	10% OV-17 Gas Chrom® Q (100/120)	300	Nitrogen (35)	NPD	4.5	*p*-Bromobenzaldehyde isonicotinoyl hydrazone (6)	a	b	1
U (10)	2	1 × 3	1.5% OV-17 Shimalite W (80/100)	250	Nitrogen (20)	FID	4	Benzoic acid hydrazide	i. c ii. Trimethylsilyl	b	2

High-Pressure Liquid Chromatography

Specimen (mℓ)	S	Column cm × mm	Packing (μm)	Elution	Flow rate (mℓ/min)	Det. (nm)	RT (min)	Internal standard (RT)	Other compounds (RT)	Ref.
D	3	100 × 2.1	HC Pellosil (NA)	E-1	1.2	UV (254)	2.3	Tribenzylamine (1.5)	Pyridoxine (4.1)	3
B,U (3)	2	30 × 3.9	μ-Bondapak® C_{18} (10)	E-2	2.0	UV (266)	5	1-Benzoyl-2-isonicotinoylhydrazine (4)	Acetylisoniazid (3)	4

Note: E-1 = Chloroform + methanol + 2-propanol + water.
85 : 5 : 10 : 0.5

E-2 = Methanol + water + dioctylsodium sulfosuccinate, pH 2.5.
60 : 40 : (4.4 g/ℓ)

[a] Reaction with *p*-chlorobenzaldehyde to form *p*-chlorobenzaldehyde isonicotinoyl hydrazone.
[b] Conditions for the quantitative analysis of hydrazine, acetylhydrazine, acetylisoniazid, and deacetylisoniazid are also described.
[c] Reaction with benzaldehyde to form benzaldehyde isonicotinoyl hydrazone.

ISONIAZID (continued)

REFERENCES

1. Timbrell, J. A., Wright, J. M., and Smith, C. M., *J. Chromatogr.*, 138, 165, 1977.
2. Noda, A., Goromaru, T., Matsuyama, K., Sogabe, K., Hsu, K.-Y., and Iguchi, S., *J. Pharm. Dyn.*, 1, 132, 1978.
3. Bailey, L. C. and Abdou, H., *J. Pharm. Sci.*, 66, 564, 1977.
4. Saxena, S. J., Stewart, J. T., Honiberg, I. L., Washington, J. G., and Keene, G. R., *J. Pharm. Sci.*, 66, 813, 1977.

ISOSORBIDE DINITRATE

Gas Chromatography

Specimen (mℓ)	S	Column m × mm	Packing (mesh)	Oven temp (°C)	Gas (mℓ/min)	Det.	RT (min)	Internal standard (RT)	Deriv.	Other compounds (RT)	Ref.
B (3)	2	1.2 × 4	10% OV-1 Chromosorb® G (80/100)	200	Argon 95 + Methane 5 (40)	ECD	3.8	—	—	Isosorbide-2 mononitrate (1.7) Isosorbide-5 mononitrate (2.6)	1
B (1)	2	1.2 × 4	3% QF-1 Gas Chrom® Q (100/120)	150	Argon 95 + Methane 5 (95)	ECD	3	Isoidide dinitrate (2)	—	—	2
B (3)	1	1.8 × 2	30% SE-30 Gas Chrom® Q (60/80)	160	NA (60)	ECD	6	—	—	Isosobide-2 mononitrate (2.6) Isosorbide-5 mononitrate (4)	3
B (3)	2	1.9 × 2	3% OV-17 Chromosorb® W (100/120)	160	Nitrogen (60)	ECD	—	*O*-Nitrobenzyl alcohol (3.5)	*O*-*t*-Butyldimethylsilyl	Isosorbide-2 mononitrate (4.5) Isosorbide-5 mononitrate (6)	4
B (2)	2	1.2 × 2.4	3% OV-101 Gas Chrom® Q (80/100)	130	Argon 95 + Methane 5 (20)	ECD	9.8	Nitroglycerin (0.8)	—	—	5
B (2)	2	25 × 0.5[a]	OV-17	165	Helium (7.5)	ECD	11.8	Isoidide mononitrate (5.7) + Isomannide dinitrate (17)	—	Isosorbide-2 mononitrate (4.2) Isosorbide-5 mononitrate (8.3)	6

[a] Wall coated open tubular column.

ISOSORBIDE DINITRATE (continued)

REFERENCES

1. Richard, L., Klein, G. and Orr, J. M., *Clin. Chem.*, 22, 2060, 1976.
2. Malbica, J. O., Monson, K., Nielson, K., and Sprissler, R., *J. Pharm. Sci.*, 66, 384, 1977.
3. Chin, D. A., Prue, D. G., Michelucci, J., Kho, B. T., and Warner, C. R., *J. Pharm. Sci.*, 66, 1143, 1977.
4. Smith, R. V. and Besic, J., *Microchem. J.*, 23, 185, 1978.
5. Laufen, H., Scharpf, F., and Bartsch, G., *J. Chromatogr.*, 146, 457, 1978.
6. Rosseel, M. T. and Bogaert, M. G., *J. Pharm. Sci.*, 68, 659, 1979.

ISOXEPAC

Gas Chromatography

Specimen (mℓ)	S	Column m × mm	Packing (mesh)	Oven temp (°C)	Gas (mℓ/min)	Det.	RT (min)	Internal standard (RT)	Deriv.	Other compounds (RT)	Ref.
B (2)	1	1.8 × 4	3% OV-11 Chromosorb® W (100/120)	265	Nitrogen (60)	FID	5.2	2-Propionic acid analog (6.9)	Methyl[a]	—	1

[a] With diazomethane.

REFERENCE

1. Bryce, T. A. and Burrows, J. L., *J. Chromatogr.*, 145, 393, 1978.

KANAMYCIN

High-Pressure Liquid Chromatography

Specimen (mℓ)	S	Column cm × mm	Packing (μm)	Elution	Flow rate (mℓ/min)	Det. (nm)	RT (min)	Internal standard (RT)	Other compounds (RT)	Ref.
D	3	100 × 2.1	HS Pellionex SCX (NA)	E-1	1	Fl[a] (Ex: 320, Fl: 450)	A[b] = 4.5 B = 8	—	—	1

Note: E-1 = 0.01 *M* Ethylenediaminetetracetic acid, pH 9.5.

[a] Post-column derivatization with *o*-phthalaldehyde reagent.
[b] Mixture of different kanamycins in fermentation-produced antibiotics.

REFERENCE

1. Mays, D. L., van Apeldoorn, R. J., and Lauback, R. G., *J. Chromatogr.*, 120, 93, 1976.

KETAMINE

Gas Chromatography

Specimen (mℓ)	S	Column m × mm	Packing (mesh)	Oven temp (°C)	Gas (mℓ/min)	Det.	RT (min)	Internal standard (RT)	Deriv.	Other compounds (RT)	Ref.
B (1)	1	1.5 × 2	3% OV-17 Gas Chrom® Q (100/120)	170	Helium (20)	MS-EI[a]	3.9	o-Br-analog (5.6)	Heptafluro-butyryl	Metabolite 1[b] (2.5) Metabolite 2[c] (2.8)	1
B (0.1—1)	1	1.8 × 2	3% SP-200 Supelcoport (100/120)	220	Nitrogen (40)	NPD[d]	2.5	Diphenylhydramine (1)	—	Metabolite 1[b] (3.5) Metabolite 2[c] (5.5)	2

High-pressure Liquid Chromatography

Specimen (mℓ)	S	Column cm × mm	Packing (μm)	Elution	Flow rate (mℓ/min)	Det. (nm)	RT (min)	Internal standard (RT)	Other compounds (RT)	Ref.
U (15)	2	30 × 4	μ-Bondapak® C_{18} (10)	E-1	1.5 UV[e]	UV[e] (254)	6.8	—	Metabolite 1[b] (8) Metabolite 2[c] (5.5)	3

Note: E-1 = Acetonitrile + water.
1 : 1

[a] MS-CI was also used.
[b] 2-(o-Chlorophenyl)-2-aminocyclohexanone.
[c] 6-(o-Chlorophenyl)-6-amino-cyclohex-2,3-en-1-one.
[d] MS-EI was also used as a detector.
[e] p-Nitrobenzamides are prepared prior to chromatography.

KETAMINE (continued)

REFERENCES

1. Lau, S. S. and Domino, E. F., *Biomed. Mass Spectrom.*, 4, 317, 1977.
2. Davisson, J. N., *J. Chromatogr.*, 146, 344, 1978.
3. Needham, L. L. and Kochhar, M. M., *J. Chromatogr.*, 114, 220, 1975.

KETAZOLAM

High-Pressure Liquid Chromatography

Specimen (mℓ)	S	Column cm × mm	Packing (μm)	Elution	Flow rate (mℓ/min)	Det. (nm)	RT (min)	Internal standard (RT)	Other compounds (RT)	Ref.
D	3	100 × NA	Corasil II (NA)	E-1	1.0	UV (254)	7.5	—	Diazepam (3.5)	1

Note: E-1 = Tetrahydrofuran + isopropyl ether.
15 : 85

REFERENCE

1. **Weber, D. J.,** *J. Pharm. Sci.,* 61, 1797, 1972.

KETOCAINE

Gas Chromatography

Specimen (mℓ)	S	Column m × mm	Packing (mesh)	Oven temp (°C)	Gas (mℓ/min)	Det.	RT (min)	Internal standard (RT)	Deriv.	Other compounds (RT)	Ref.
B(0.2)	1	2.1 × 2	10% Carbowax® 20*M* + 1% KOH Chromosorb® W (60/80)	235	Helium (25)	MS-EI	2.5	2′-(2-diisopropylamino-ethoxy) caprophenone (4.6)	—	—	1

REFERENCE

1. Berlin-Wahlen, A., Hogberg, K., Lindgren, J. E., and Vecchietti, M., *Acta Pharm. Suec.*, 14, 425, 1977.

KETOPROFEN

Gas Chromatography

Specimen (mℓ)	S	Column m × mm	Packing (mesh)	Oven temp (°C)	Gas (mℓ/min)	Det.	RT (min)	Internal standard (RT)	Deriv.	Other compounds (RT)	Ref.
B (1)	2	2 × 3	3% OV-17 Chromosorb® W (NA)	240	Nitrogen (30)	FID	5	2-(3-Benzoyl phenyl) butyric acid (6.8)	Methyl[a,b]	—	1

High-Pressure Liquid Chromatography

Specimen (mℓ)	S	Column cm × mm	Packing (μm)	Elution	Flow rate (mℓ/min)	Det. (nm)	RT (min)	Internal standard (RT)	Other compounds (RT)	Ref.
B (1)	2	25 × 4.7	LiChrosorb® Si-60 (5)	E-1	1.3	UV (254)	12.8[c]	2-(3-Benzoyl-phenyl) propionic acid (11.6)[c]	—	1
B, U (0.5)	2	30 × 4	μ-Bondapak® C_{18} (10)	E-2	1.1	UV (254)	5	2-(3-Benzoyl-phenyl) propionic acid (8)	—	2
B, U (2)	2	50 × 4.6	Spherisorb-5 ODS (5)	E-3	2	UV (260)	3	Oxyphenbutazone (2)	—	3

Thin-Layer Chromatography

Specimen (mℓ)	S	Plate (manufacturer)	Layer (mm)	Solvent	Post-separation treatment	Det. (nm)	R_f	Internal standard (R_f)	Other compounds (R_f)	Ref.
B (0.25)	1	NA (Merck)	Silica gel 60 F_{254} (0.25)	S-1	—	UV[d] (255)	0.69	—	—	4

KETOPROFEN (continued)

Note: E-1 = Dichloromethane + hexane.
60 : 40

E-2 = Methanol + water.
45 : 55

E-3 = Methanol + water + (acetic acid, pH 3.5).
35 : 65

S-1 = Ether + benzene + 1-butanol + methanol.
85 : 8 : 6 : 1

[a] With diazomethane.
[b] Methyl esters are purified by TLC on silica gel plates using hexane-acetone (85:15) as the developing solvent, prior to analysis by GC.
[c] The acids are derivatized with diazomethane prior to analysis by liquid chromatography.
[d] The spots are scraped, eluted, and the absorbance of the supernatants determined spectrophotometrically @ 255 nm.

REFERENCES

1. **Bannier, A., Brazier, J. L., and Ribon, B.,** *J. Chromatogr.,* 155, 371, 1978.
2. **Farinotti, R. and Mahuzier, G.,** *J. Pharm. Sci.,* 68, 484, 1979.
3. **Jefferies, T. M., Thomas, W. O. A., and Parfitt, R. T.,** *J. Chromatogr.,* 162, 122, 1979.
4. **Ballerini, R., Cambi, A., and Soldato, P. D.,** *J. Pharm. Sci.,* 66, 281, 1977.

LABETALOL

High-Pressure Liquid Chromatography

Specimen (mℓ)	S	Column cm × mm	Packing (μm)	Elution	Flow rate (mℓ/min)	Det. (nm)	RT (min)	Internal standard (RT)	Other compounds (RT)	Ref.
B (2)	2	30 × 4	μ-Bondapak® C_{18} (10)	E-1	1.0	UV (233)	3.9	Pericyazine (7.3)	—	1

Note: E-1 = Acetonitrile + 45 m*M* KH_2PO_4, pH 3.
35 : 65

REFERENCE

1. **Dusci, L. J. and Hacket, L. P.,** *J. Chromatogr.*, 175, 208, 1979.

LASALOCID

Gas Chromatography

Specimen (mℓ)	S	Column m × mm	Packing (mesh)	Oven temp (°C)	Gas (mℓ/min)	Det.	RT (min)	Internal standard (RT)	Deriv.	Other compounds (RT)	Ref.
D	3	2 × 3	10% OV-17 Gas Chrom® Q (100/120)	250	Helium (70)	FID	—	Methylbehenate (20)	—	Retroaldolketone[a] (12)	1

[a] Formed by pyrolytic cleavage of lasalocid.

REFERENCE

1. Westley, J. W., Evans, R. H., Jr., and Stempel, A., *Anal. Biochem.*, 59, 574, 1974.

LIDOCAINE

Gas Chromatography

Specimen (mℓ)	S	Column m × mm	Packing (mesh)	Oven temp (°C)	Gas (mℓ/min)	Det.	RT (min)	Internal standard (RT)	Deriv.	Other compounds (RT)	Ref.
B (2.5)	2	1.8 × 3	2% Carbowax® 20 *M* + 3% KOH Chromosorb® W (80/100)	210	Nitrogen (15)	NPD	4.5	Pyrrocaine (7.5)	—	Monoethylglyci-nexylidide (6)	1
B (1)	2	1.8×2	3% OV-17 Gas Chrom® Q (100/120)	200	Helium (50)	NPD	3.5	Ethylmethylgly-cinexylidide (2.5)	—	—	2
B (1)	2	0.75 × NA	2% Ucon® 75-H-90,000 + 2% KOH (80/100)	175	Nitrogen (50)	FID	4.5	Benzhexol (10)	—	Monoethylglyci-nexylidide (7.6)	3
B (2)	2	1.8 × NA	3% OV-17 Gas Chrom® Q (80/100)	210	Nitrogen (56)	FID	4.5	Aminopyrine (6.7)	—	—	4
B (1)	2	1.8 × 2	3% OV-17 Gas Chrom® Q (100/120)	210	Nitrogen (30)	FID	1.2	Mepivacaine (3.5)	—	—	5
B (2)	2	20 × 0.5[a]	OV-17	190	Hydrogen (4)	NPD	4	Trimecaine (5.5)	Trifluo-roacetyl	Colycine-xylidide (3) Monoethylglyci-nexylidide (3.5)	6
Tissue	2	2.8 × 4	3% OV-101 Supelcoport (80/100)	195	Nitrogen (60)	FID	6.5	Mianserin (18)	—	—	7

LIDOCAINE (continued)

High-Pressure Liquid Chromatography

Specimen (mℓ)	S	Column cm × mm	Packing (μm)	Elution	Flow rate (mℓ/min)	Det. (nm)	RT (min)	Internal standard (RT)	Other compounds (RT)	Ref.
B (1)	2	50 × 2.6	Sil-X-1[b] ODS (NA)	E-1	2	UV (205)	7.8	Procaine (5.8)	Procainamide (4)	8
B (0.5)	2	25 × 2.6	Sil-X-1[b] ODS (NA)	E-2	1	UV (205)	3.2	Bupivicaine (7.5)	Thiopental (5.9)	9
B (0.5)	2	30 × 3.9	μ-Bondapak® C_{18} (10)	E-3	2	UV (210)	9.3	Ethylmethyl glycinexylidide (7.3)	Glycinexylidide (4.3) Monoethylglycinexylidide (6.1)	10
B (0.5)	2	30 × 3.9	μ-Bondapak® alkyl phenyl (10)	E-4	2 (200)	UV (200)	10	Ethylmethyl glycinexylidide (8.3)	Glycinexylidide (5.9) Monoethylglycinexylidide (7.4)	11
B (NA)	2	30 × 4	μ-Bondapak®[c] C_{18} (10)	E-5	2.0	UV (205)	20	—	Glycinexylidide (7) Monoethylglycinexylidide (10.5)	12

Thin-Layer Chromatography

Specimen (mℓ)	S	Plate (manufacturer)	Layer (mm)	Solvent	Post-separation treatment	Det. (nm)	R_f	Internal standard (R_f)	Other compounds (R_f)	Ref.
B (0.5)	2	10 × 10 cm (Merck)[d]	Silica gel 60 F_{254} (NA)	S-1	—	UV (220)	0.34	Clozapine (0.09)	e	13

Note: E-1 = Acetonitrile + 0.2 *M* phosphate buffer, pH 6.0.
10 : 90

E-2 = Acetonitrile + 0.2 *M* phosphate buffer, pH 4.0.
10 : 90

E-3 = Acetonitrile + 0.1 *M* phosphate buffer, pH 4.0.
8 : 92

E-4 = Acetonitrile + 0.006% phosphoric acid.
30 : 70

E-5 = Acetonitrile + 0.05 *M* KH_2PO_4.
5 : 95

S-1 = Benzene + ethyl acetate + methanol.
4 : 4 : 1

[a] Wall-coated open tubular column.
[b] Column temperature = 40°C.
[c] Column is protected by a precolumn packed with CoPell ODS.
[d] High performance thin layer plates.
[e] Conditions for the estimation of other antiarrhythmic drugs have been described.

REFERENCES

1. Irgens, T. R., Henderson, W. M., and Shelver, W. H., *J. Pharm. Sci.*, 65, 608, 1976.
2. Hucker, H. B. and Stauffer, S. C., *J. Pharm. Sci.*, 65, 926, 1976.
3. Nation, R. L., Triggs, E. J., and Selig, M., *J. Chromatogr.*, 116, 188, 1976.
4. Caille, G., Lelorier, J., Latour, Y., and Besner, J. G., *J. Pharm. Sci.*, 66, 1383, 1977.
5. Kline, B. J. and Martin, M. F., *J. Pharm. Sci.*, 67, 887, 1978.
6. Rosseel, M. T. and Bogaert, M. G., *J. Chromatogr.*, 154, 99, 1978.
7. Holt, D. W., Loizou, M., and Wyse, R. K. H., *J. Clin. Path.*, 32, 225, 1979.
8. Adams, R. F., Vandemark, F. L., and Schmidt, G., *Clin. Chim. Acta*, 69, 515, 1976.
9. Masoud, A. N., Scratchley, G. A., Stohs, S. J., and Wingard, D. W., *J. Liq. Chromatogr.*, 1, 607, 1978.
10. Naranang, P. K., Crouthamel, W. G., Carliner, N. H., and Fisher, M. L., *Clin. Pharmacol. Ther.*, 24, 654, 1978.

LIDOCAINE (continued)

11. **Nation, R. L., Peng, G. W., and Chiou, W. L.,** *J. Chromatogr.*, 162, 466, 1979.
12. **Wisnicki, J. L., Tong, W. P., and Ludlum, D. B.,** *Clin. Chim. Acta*, 93, 279, 1979.
13. **Lee, K. Y., Nurok, D., Zlatkis, A., and Karman, A.,** *J. Chromatogr.*, 158, 403, 1978.

LOFEPRAMINE

Gas Chromatography

Specimen (mℓ)	S	Column m × mm	Packing (mesh)	Oven temp (°C)	Gas (mℓ/min)	Det.	RT (min)	Internal standard (RT)	Deriv.	Other compounds (RT)	Ref.
B (2)	2	1.5 × 3	3% HI-EFF-8BP Gas Chrom® Q (80/100)	150	Nitrogen[a]	ECD	1.5	Bromo analog (2.5)	b	c	1
B (1)	1	2 × 3	2% PEG-20 *M* Gas Chrom® Q (80/100)	220	Helium (30)	MS-EI	2.5	$^2H_4$-*p*-Chloro-benzoic acid (2.5)	d	c	2

[a] Inlet pressure = 3.9×10^5 N/m². (N/m¹ = newtons/square meter, SI unit for pressure.)
[b] Reduction with sodium borohydride; oxidation with sodium metaperiodate and distillation to isolate *p*-chloro and *p*-bromo benzaldehydes.
[c] Conditions for the estimation of desipramine are described.
[d] Oxidation with hydrogn peroxide to *p*-chloro benzoic acid and preparation of ethyl esters of *p*-chloro benzoic acid and of deuterated *p*-chloro benzoic acid with ethanol and HC1.

REFERENCES

1. Lundgreen, R., Olsson, A., and Forshell, G. P., *Acta Pharm. Suec.*, 14, 81, 1977.
2. Matsubayashi, K., Hakusui, H., and Sano, M., *J. Chromatogr.*, 143, 571, 1977.

LORAZEPAM

Gas Chromatography

Specimen (mℓ)	S	Column m × mm	Packing (mesh)	Oven temp (°C)	Gas (mℓ/min)	Det.	RT (min)	Internal standard (RT)	Deriv.	Other compounds (RT)	Ref.
B,U (1)	1	1.8 × 4	3% OV-17 Chromosorb® W (80/100)	280	Helium (50)	ECD	2.7[a]	Oxazepam (2.4)[a]	—	—	1

[a] It is presumed that lorazepam and oxazepam decompose on column to produce quinazoline carboxaldehyde derivatives.

REFERENCE

1. Greenblatt, D. J., Franke, K., and Shader, R. I., *J. Chromatogr.*, 146, 311, 1978.

LORCAINIDE

Gas Chromatography

Specimen (mℓ)	S	Column m × mm	Packing (mesh)	Oven temp (°C)	Gas (mℓ/min)	Det.	RT (min)	Internal standard (RT)	Deriv.	Other compounds (RT)	Ref.
B (1)	1	2 × 2	3% OV-22 Supelcoport (80/100)	260	Nitrogen (40)	ECD	2.5	*n*-[1-(3-Methylbutyl)-analog (3.3)	—	Norlorcainide (1.9)	1
										4-Hydroxylorcainide (4.3)	
										4-Hydroxy-3-methoxylorcainide (5.9)	

REFERENCE

1. Woestenborghs, R., Michiels, M., and Heykants, J., *J. Chromatogr.*, 164, 169, 1979.

LOXAPINE

Gas Chromatography

Specimen (mℓ)	S	Column m × mm	Packing (mesh)	Oven temp (°C)	Gas (mℓ/min)	Det.	RT (min)	Internal Standard (RT)	Deriv.	Other compounds (RT)	Ref.
B (1—3)	2	1.8 × 2	3% SP-2100 Supelcoport (100/120)	255	Argon 95 + Methane 5 (NA)	ECD	4	8-Methoxyloxapine (8)	i. Trifluoroacetyl ii. Trimethylsilyl	Loxapine-*N*-oxide (5.5) 8-Hydroxyloxapine (9) 7-Hydroxyloxapine (8.5) Amoxapine 8-Hydroxyamoxapine (13)	1

REFERENCE

1. Cooper, T. B. and Kelly, R. G., *J. Pharm. Sci.*, 68, 216, 1979.

LYSERGIDE

Gas Chromatography

Specimen (mℓ)	S	Column m × mm	Packing (mesh)	Oven temp (°C)	Gas (mℓ/min)	Det.	RT (min)	Internal standard (RT)	Deriv.	Other compounds (RT)	Ref.
D	3	1.8 × 2	0.25% OV-17 GLC-110 (textured glass beads) (100/120)	258	NA (30)	FID	7.5	—	Trimethyl-silyl	—	1

High-Pressure Liquid Chromatography

Specimen (mℓ)	S	Column cm × mm	Packing (μm)	Elution	Flow rate (mℓ/min)	Det. (nm)	RT (min)	Internal standard (RT)	Other compounds (RT)	Ref.
D	3	50 × 2.3	Sil-X[a] (NA)	E-1	1	UV (254)	11.5	—	Iso-LSD (32.4) Phencyclidine (3.8) Ergotamine (9.6)	2
U (40)	3	25 × 4.6	Partisil® (6)	E-2	1	Fl (Ex: 325, Fl: 430)	9	—	Iso-LSD (13)	3
B,U (NA)	2	10 × 4.6	Spherisorb® 5[b] ODS (5)	E-3	1	Fl (Ex: 320, Fl: 400)	4.9	—	Iso-LSD[c] (10)	4

Thin-Layer Chromatography

Specimen (mℓ)	S	Plate (manufacturer)	Layer (mm)	Solvent	Post-separation treatment	Det. (nm)	R_f	Internal standard (R_f)	Other compounds (R_f)	Ref.
U (5)	3	20 × 20 cm	Filter paper No.	S-1	—	i. Visual (UV-	0.79	—	—	5

LYSERGIDE (continued)

Thin-Layer Chromatography

Specimen (mℓ)	S	Plate (manufacturer)	Layer (mm)	Solvent	Post-separation treatment	Det. (nm)	R_f	Internal standard (R_f)	Other compounds (R_f)	Ref.
		(Whatman)	1			254) ii. Fl[d]				
D	3	10 × 10 cm (Analtech)	Silica gel G F^{254} (0.25)	S-2	Sp: 2% Dimethylaminobenzaldehyde in 50% HCl in methanol	i. Visual (UV-254) ii. Visual after spray	0.55	—	e	1
D	3	NA (Merck)	Silica gel G (0.25)	S-3	—	UV reflectance (310)	0.55	—	2,5-Dimethoxy-4-methylamphetamine (0.15)	6

Note: E-1 = Acetonitrile + isopropyl ether.
40 : 60

E-2 = Methanol + 0.2% ammonium nitrate.
11 : 9

E-3 = Methanol + 0.025 *M* Na_2HPO_4, pH 8.0.
65 : 35

S-1 = 1-Butanol + acetic acid (saturated with water).
10 : 1

S-2 = Chloroform (saturated with ammonia) + methanol.
18 : 1

S-3 = Acetone + chloroform + methanol.
3 : 1 : 1

[a] Alternatively, 60 cm × 2.3 mm column packed with Corasil-II was also used.
[b] A normal phase column (15 cm × 4.6 mm packed with Spherisorb® S5W) was also used.
[c] Retention data for a number of other ergot alkaloids are given.

[d] The spots are cut and diluted with dilute HCl. A portion of the eluate is irradiated with UV (254 nm) light for 30 min and the fluorescence of both irradiated samples is determined.

[e] R_f values of a number of ergot alkaloids are given.

REFERENCES

1. Sperling, A. R., *J. Chromatogr. Sci.*, 12, 265, 1974.
2. Wittwer, J. D., Jr., and Kluckhohn, J. H., *J. Chromatogr. Sci.*, 11, 1, 1973.
3. Christie, J., White, M. W., and Wiles, J. M., *J. Chromatogr.*, 120, 496, 1976.
4. Twitchett, P. J., Fletcher, S. M., Sullivan, A. T., and Moffat, A. C., *J. Chromatogr.*, 150, 73, 1978.
5. Faed, E. M. and McLeod, W. R., *J. Chromatogr. Sci.*, 11, 4, 1973.
6. Niwaguchi, T. and Inoue, T., *J. Chromatogr.*, 121, 165, 1976.

MAFENIDE

Thin-Layer Chromatography

Specimen (mℓ)	S	Plate (manufacturer)	Layer (mm)	Solvent	Post-separation treatment	Det. (nm)	R_f	Internal standard (R_f)	Other compounds (R_f)	Ref.
B (0.05)	2	10 × 20 cm (Merck)	Silica gel 60 (NA)	S-1	Sp: Fluorescamine (15 mg) in acetone (200 mℓ)	Fl (Ex: 390, Fl: 490)	NA	—	—	1

Note: S-1 = Ethyl acetate + methanol + ammonia.
75 : 20 : 5

REFERENCE

1. **Steyn, J. M.,** *J. Chromatogr.*, 143, 210, 1977.

MAPROTILINE

Gas Chromatography

Specimen (mℓ)	S	Column m × mm	Packing (mesh)	Oven temp (°C)	Gas (mℓ/min)	Det.	RT (min)	Internal standard (RT)	Deriv.	Other compounds (RT)	Ref.
B (1)	2	1.2 × 3	3% JXR Gas Chrom® Q (NA)	230	Nitrogen (40)	ECD	4.5	Nortriptyline (3)	Heptafluorobutyryl	—	1
B (2—3)	2	1.2 × 2	3% HI-EFF-8BP Gas Chrom® Q (80/100)	240	Nitrogen (25)	NPD	12.3	Desmethyldoxepin (9.5)	Acetyl	Desmethylmaprotiline (19.1)	2

REFERENCES

1. Geiger, U. P., Rajagapalan, T. G., and Riess, W., *J. Chromatogr.*, 114, 167, 1975.
2. Gupta, R. N., Molnar, G., and Gupta, M. L., *Clin. Chem.*, 23, 1849, 1977.

MARIDOMYCIN

High-Pressure Liquid Chromatography

Specimen (mℓ)	S	Column cm × mm	Packing (μm)	Elution	Flow rate (mℓ/min)	Det. (nm)	RT (min)	Internal standard (RT)	Other compounds (RT)	Ref.
D	3	200 × 2	Corasil I (37—50)	E-1	0.35	Refractive index	a 1 = 20 2 = 23 3 = 26 4 = 30 5 = 35 6 = 43	Neopentylglycol (14)	b	1

Thin-Layer Chromatography

Specimen (mℓ)	S	Plate (manufacturer)	Layer (mm)	Solvent	Post-separation treatment	Det. (nm)	R_f	Internal standard (R_f)	Other compounds (R_f)	Ref.
D	3	20 × 20 cm (Laboratory)	Silica gel G-60 (0.25)	S-1	E: I_2 vapors for 30 min	c	—	—	9-Propionylmari-domycins 1 = 0.6 2 = 0.55 3 = 0.48 4 = 0.4 5 = 0.35 6 = 0.3	2

Note: E-1 = (n-Hexane + diisopropyl ether + ethanol + water) upper layer + ethanol
1 : 4 : 1 : 2 96 : 4

S-1 = n-Hexane + diisopropyl ether + ethanol + water (upper layer)
3 : 20 : 5 : 4

[a] Components of maridomycin isolated from the culture medium of streptomyces hydroscopius No. B-5050.
[b] Chromatographic conditions and retention data for 9-propionyl derivatives of maridomycin components is given.
[c] The spots were scraped, eluted with 0.1 N sodium hydroxide and iodine content of each fraction was determined.

REFERENCES

1. **Kondo, K.**, *J. Chromatogr.*, 169, 329, 1979.
2. **Kondo, K.**, *J. Chromatogr.*, 169, 337, 1979.

MEBENDAZOLE

High-Pressure Liquid Chromatography

Specimen (mℓ)	S	Column cm × mm	Packing (μm)	Elution	Flow rate (mℓ/min)	Det. (nm)	RT (min)	Internal standard (RT)	Other compounds (RT)	Ref.
B (2)	2	30 × 4	μ-Bondapak® C_{18} (10)	E-1	2.5	UV (313)	10	Flubendazole (12.6)	2-Amino-5-benzoylbenzimidazole (5.6) 5-Benzoyl-2-hydroxybenzimidazole (4.9) Methyl-5-(α-hydroxybenzyl)-2-benzimidazole carbamate (4.9)	1

Note: E-1 = Acetonitrile + 0.05 *M* phosphate buffer, pH 6.0.
27 : 73

REFERENCE

1. **Alton, K. B., Patrick, J. E., and McGuire, J. L.,** *J. Pharm. Sci.*, 68, 880, 1979.

MECLIZINE

Gas Chromatography

Specimen (mℓ)	S	Column m × mm	Packing (mesh)	Oven temp (°C)	Gas (mℓ/min)	Det.	RT (min)	Internal standard (RT)	Deriv.	Other compounds (RT)	Ref.
D	3	1.8 × 4 Steel	3% OV-17 Chromosorb® W (80/100)	290	Helium (75)	FID	14	Dinonylphthalate (7)	—	—	1
D	3	1.5 × 1.8	3% DC560 + 3% NPGSE Gas Chrom® P (NA)	145	Nitrogen (30)	ECD	NA	—	a	b	2
B (4)	1	0.8 × 2	3% OV-17 Chromosorb® W (80/100)	285	Methane[c]	MS-CI	NA	[$^2H_5$]-Meclizine (NA)	—	—	3

[a] The drug is oxidized in quantitative yield to benzophenone with barium peroxide in 0.5 *M* sulfuric acid in 90 min.

[b] A large number of drugs with a gem-diphenylmethane group have been oxidized to benzophenone.

[c] Ammonia into ion source to give a pressure, 0.35 to 0.4 Torr; methane flow through the column was adjusted to give a final pressure of 0.95 to 1.0 Torr.

REFERENCES

1. Wong, C. K., Urbigkit, J. R., Conca, N., Cohen, D. M., and Munnely, K. P., *J. Pharm. Sci.*, 62, 1340, 1973.
2. Hartvig, P. and Handl, W., *Acta Pharm. Suec.*, 12, 349, 1975.
3. Fouda, H. G., Falkner, F. C., Hobbs, D. C., and Luther, E. W., *Biomed. Mass Spectrom*, 5, 491, 1978.

MEDAZEPAM

Gas Chromatography

Specimen (mℓ)	S	Column m × mm	Packing (mesh)	Oven temp (°C)	Gas (mℓ/min)	Det.	RT (min)	Internal standard (RT)	Deriv.	Other compounds (RT)	Ref.
B, U (4)	2	1.5 × 3	3% OV-25 Chromosorb® W (80/100)	Temp. Progr.	Helium (60)	NPD	2.5	2-Methylamino-5-chlorobenzophenone (1.7)	—	*N*-Desmethylmedazepam (3.4) Diazepam (4.1)	1
B (25)	2	1.8 × 4	3% OV-17 Gas Chrom® Q (60/80)	Temp. Progr.	Argon (50)	[a]	7	—	—	*N*-Desmethylmedazepam (9) Diazepam (13)	2
B (5)	2	1.8 × 2	3% OV-17 Gas Chrom® Q (60/80)	Temp. Progr.	Helium (25)	[a]	NA	—	—	*N*-Desmethylmedazepam (NA)	3

[a] Electrolytic conductivity.

REFERENCES

1. **Mallach, H. J., Moosmayer, A., and Rupp, J. M.,** *Arzneim. Forsch.*, 23, 614, 1973.
2. **Hailey, D. M., Howard, A. G., and Nickless, G .,** *J. Chromatogr.*, 100, 49, 1974.
3. **Pape, B. E. and Ribick, M. A.,** *J. Chromatogr.*, 136, 127, 1977.

MEFENAMIC ACID

Gas Chromatography

Specimen (mℓ)	S	Column m × mm	Packing (mesh)	Oven temp (°C)	Gas (mℓ/min)	Det.	RT (min)	Internal standard (RT)	Deriv.	Other compounds (RT)	Ref.
B,U (2)	2	1.2 × 2	3% OV-1 Chromosorb G (80/100)	190	Helium (40)	ECD	3.7	—	Pentafluoropropyl	—	1

Thin-Layer Chromatography

Specimen (mℓ)	S	Plate (manufacturer)	Layer (mm)	Solvent	Post-separation treatment	Det. (nm)	R_f	Internal standard (R_f)	Other compounds (R_f)	Ref.
U (10)	3	NA (Laboratory)	Silica gel G (0.25)	S-1	Sp: 1% sodium nitrite in 1% sulfuric acid	Visual	0.75	—	Metabolite A[a] (0.45) Metabolite B[b] (0.28) Flufenamic acid (0.73) Meclofenamic acid (0.74)	2

Note: S-1 = Toluene + acetic acid.
9 : 1

[a] N-(2-methyl-3-carboxyphenyl) anthranilic acid.
[b] N-(2-methyl-3-hydroxymethylphenyl) anthranilic acid.

MEFENAMIC ACID (continued)

REFERENCES

1. **Bland, S. A., Blake, J. W., and Ray, R. S.,** *J. Chromatogr. Sci.*, 14, 201, 1976.
2. **Demetriou, B. and Osborne, B. G.,** *J. Chromatogr.*, 90, 405, 1974.

MEFLOQUINE

Gas Chromatography

Specimen (mℓ)	S	Column m × mm	Packing (mesh)	Oven temp (°C)	Gas (mℓ/min)	Det.	RT (min)	Internal standard (RT)	Deriv.	Other compounds (RT)	Ref.
B (5)	1	1.8 × 4	3% OV-17 Chromosorb® W (100/120)	Temp. Progr.	Nitrogen (30)	ECD	NA	α-2′-Piperidyl-2-(4-trifluoromethylphenyl)-6-trifluoromethyl-4-pyridine methanol (NA)	Trimethylsilyl	—	1

High-Pressure Liquid Chromatography

Specimen (mℓ)	S	Column cm × mm	Packing (μm)	Elution	Flow rate (mℓ/min)	Det. (nm)	RT (min)	Internal standard (RT)	Other compounds (RT)	Ref.
B, U (5)	1	30 × 4	μ-Bondapak® CN[a] (10)	E-1	2	UV (280)	4	2,8-Bis(trifluoromethyl)-4-[1-hydroxy-3-(N-tertbutylamino) propyl] quinoline (6)	—	2

Note: E-1 = Isopropyl ether + dioxane + acetic acid.
60 : 40 : 0.5

[a] A μ-Bondapak® C_{18} column (30 cm × 4 mm, i.d.) was used for the analysis of urine extracts using a mobile phase of methanol—0.1 *M* NaH_2PO_4 (3:2) at 2 mℓ/min.

MEFLOQUINE (continued)

REFERENCES

1. Nakagawa, T., Higuchi, T., Haslam, J. L., Shaffer, R. D., and Mendenhall, D. W., *J. Pharm. Sci.*, 68, 718, 1979.
2. Grindel, J. M., Tilton, P. F., and Shaffer, R. D., *J. Pharm. Sci.*, 66, 834, 1977.

MEFRUSIDE

Gas Chromatography

Specimen (mℓ)	S	Column m × mm	Packing (mesh)	Oven temp (°C)	Gas (mℓ/min)	Det.	RT (min)	Internal standard (RT)	Deriv.	Other compounds (RT)	Ref.
B (1)	1	1 × 2.2	5% Dexsil® 300-GC Gas Chrom® Q (80/100)	340	Helium (55)	NPD	3	—	Methyl[a]	Oxomefruside (4.5)	1
B (2)	1	1.8 × 3	3% SE-30 Gas Chrom® Q (100/120)	265	Helium (55)	NPD	4.1	4-Chloro-*N*-1-methyl-N'-(3-methoxypropyl)-1,3-benzenedisulfonamide (2.8)	Methyl[b]	—	2

High-Pressure Liquid Chromatography

Specimen (mℓ)	S	Column cm × mm	Packing (μm)	Elution	Flow rate (mℓ/min)	Det. (nm)	RT (min)	Internal standard (RT)	Other compounds (RT)	Ref.
U (10)	2	50 × NA	LiChrosorb® Si-60 (5)	E-1	1.5	UV (248)	10.5	—	Oxomefruside (22)	3

Note: E-1 = Hexane + ethyl acetate.
1 : 1

[a] Dimethylformamide — dimethylacetal reagent.
[b] Extractive alkylation with methyl iodide and tetrahexylammonium hydroxide.

MEFRUSIDE (continued)

REFERENCES

1. Oesterhelt, G. and Eschenhof, E., *Arzneim. Forsch.*, 29, 607, 1979.
2. Fleuren, H. L. J., Verwey-van Wissen, C. P. W., and van Rossum, J. M., *Arzneim. Forsch.*, 29, 1041, 1979.
3. Little, C. J., Dale, A. D., Ord, D. A., and Marten, T. R., *Anal. Chem.*, 49, 1311, 1977.

MELPHALAN

Gas Chromatography

Specimen (mℓ)	S	Column m × mm	Packing (mesh)	Oven temp (°C)	Gas (mℓ/min)	Det.	RT (min)	Internal standard (RT)	Deriv.	Other compounds (RT)	Ref.
D	3	1.8 × 3	2.5% SE-54 Chromosorb® W (80/100)	210	Nitrogen (30)	FID	27	Chrysene (18)	Trimethylsilyl	—	1

High-Pressure Liquid Chromatography

Specimen (mℓ)	S	Column cm × mm	Packing (μm)	Elution	Flow rate (mℓ/min)	Det. (nm)	RT (min)	Internal standard (RT)	Other compounds (RT)	Ref.
B (NA) Tissue	1	30 × 4	μ-Bondapak® C_{18} (10)	E-1 (gradient)	1.5	UV (254)	9.2	—	Monohydroxymelphalan (3.8) Dihydroxymelphalan (2)	2
B (1)	1	30 × 4	μ-Bondapak® C_{18} (10)	E-2	2	UV (254)	4	Dansylproline (8)	—	3

Note: E-1 = (i) 2-Methoxyethanol + 0.1% acetic acid; (ii) 2-Methoxyethanol + 0.1% acetic acid.
30 : 70 55 : 45

E-2 = Methanol + water + acetic acid.
50 : 50 : 1

MELPHALAN (continued)

REFERENCES

1. Goras, J., Knight, J. B., Iwamoto, R. H., and Lim, P., *J. Pharm. Sci.*, 59, 561, 1970.
2. Furner, R. L., Mellet, L. B., Brown, R. K., and Duncan, G., *Drug. Metab. Dispos.*, 4, 577, 1976.
3. Chang, S. Y., Alberts, D. S., Melnick, L. R., Walson, P. D., and Salmon, S. E., *J. Pharm. Sci.*, 67, 679, 1978.

MEPERIDINE

Gas Chromatography

Specimen (mℓ)	S	Column m × mm	Packing (mesh)	Oven temp (°C)	Gas (mℓ/min)	Det.	RT (min)	Internal standard (RT)	Deriv.	Other compounds (RT)	Ref.
B (5) U (10)	2	1.8 × 2	3% OV-7 Chromosorb® W (80/100)	250	Helium (40)	FID	2.2	—	—	Normeperidine (2.5)	1
B (1)	2	1.5 × 3	3% OV-17 Gas Chrom® Q (100/120)	180	Helium (22)	FID	3.8	*N*-Methyl-*N*-ethyl lidocaine (5.1)	—	—	2
B (5)	1	1.2 × NA	8% Carbowax® + 2% KOH Chromosorb® W (80/100)	195	Nitrogen (40)	FID	6.0	Benzamphetamine (7.3)	—	Norpethidine (10.2)	3
B (0.5—2)	1	1.8 × 2	3% OV-17 Gas Chrom® Q (80/100)	175	Helium (34)	FID	1.8	2-Ethyl-5-methyl-3,3-diphenyl-1-pyrroline (4.2)	Heptafluorobutyryl	Normeperidine (2.8)	4
B (1)	1	2 × 2	3% SP 2250 Chromosorb® W (80/100)	215	Nitrogen (20)	FID	1.4	N,N′-Diethylaminoacetyl-2,6-xylidine (2.5)	—	—	5
B (1)	1	0.35 × 1.5	5% SE-52 Gas Chrom® Q (100/120)	160	Helium (25)	MS-EI	0.7	$^2H_4$-Pethidine (0.7) + $^2H_5$-Norpethidine (1.9)	Trifluoroacetyl	Norpethidine (1.9)	6
U (1)	2	0.9 × 2	3% OV-17 Gas Chrom® Q (80/100)	Temp. Progr.	Helium (30)	MS-EI	—	$^2H_4$-Pethidinic acid (1.5) + $^2H_5$-Norpethidinic acid (6.1)	Pentafluorobenzyl	Pethidinic acid (1.5) Norpethidinic acid (6.1)	7
B,U	3	2.4 × 2	10% OV-1	Temp.	Helium	FID	4.2	Mepivacaine	—	Normepericine	8

MEPERIDINE (continued)

Gas Chromatography

Specimen (mℓ)	S	Column m × mm	Packing (mesh)	Oven temp (°C)	Gas (mℓ/min)	Det.	RT (min)	Internal standard (RT)	Deriv.	Other compounds (RT)	Ref.
(5—10) Tissue			NA (80/100)	Progr.	(30)			(8.2)		(4.8)	

REFERENCES

1. Knowles, J. A., White, G. R., and Ruelius, H. W., *Anal. Lett.*, 6, 281, 1973.
2. Mather, L. E. and Tucker, G. T., *J. Pharm. Sci.*, 63, 306, 1974.
3. Chan, K., Kendall, M. J. and Mitchard, M., *J. Chromatogr.*, 89, 169, 1974.
4. Szeto, H. H. and Inturrisi, C. E., *J. Chromatogr.*, 125, 503, 1976.
5. Evans, M. A. and Harbison, R. D., *J. Pharm. Sci.*, 66, 599, 1977.
6. Lindberg, C., Berg, M., Boreus, L. O., Hartvig, P., Karlsson, K.-E., Palmer, L., and Thornblad, A.-M., *Biomed. Mass Spectrom.*, 5, 540, 1978.
7. Lindberg, C., Karlsson, K.-E., and Hartvig, P., *Acta Pharm. Suec.*, 15, 327, 1978.
8. Siek, T. J., *J. Forensic Sci.*, 23, 6, 1978.

MEPHENYTOIN

Gas Chromatography

Specimen (mℓ)	S	Column m × mm	Packing (mesh)	Oven temp (°C)	Gas (mℓ/min)	Det.	RT (min)	Internal standard (RT)	Deriv.	Other compounds (RT)	Ref.
B (1)	2	1.8 × 2	3% OV-225 Gas Chrom® Q (100/120)	230	Helium (45)	FID	0.7	5-Methyl-5-phenylhydantoin (2.8)	Pentyl[a]	Desmethyl mephenytoin (3.2)	1
B (1)	1	1.2 × 2	2% OV-101 Chromosorb® W (80/100)	175	Helium (30)	MS-EI	NA	5-Methyl-5-phenylhydantoin + 3-Methyl-5-cyclopropyl-5-phenylhydantion (NA)	Ethyl[b]	Desmethyl mephenytoin (NA)	2

[a] Greely's procedure (*Clin. Chem.*, 20, 192, 1979).
[b] With ethyl iodide in the presence of KOH.

REFERENCES

1. Raisys, V. A., Zebelman, A. M., and MacMilan, S. F., *Clin. Chem.*, 25, 172, 1979.
2. Yonekawa, W. and Kupferberg, H. J., *J. Chromatogr.*, 163, 161, 1979.

MEPHOBARBITAL
Gas Chromatography

Specimen (mℓ)	S	Column m × mm	Packing (mesh)	Oven temp (°C)	Gas (mℓ/min)	Det.	RT (min)	Internal standard (RT)	Deriv.	Other compounds (RT)	Ref.
B (1)	2	1.8 × 2	3% OV-101 Chromosorb® G (100/120)	195	Nitrogen (50)	FID	4.4	Alphenal (9.8)	Butyl[a]	Phenobarbital (8.3)	1

[a] On-column butylation with tetrabutylammonium hydroxide.

REFERENCE

1. Hooper, W. D., Dubetz, D. K., Eadie, M. J., and Tyrer, J. H., *J. Chromatogr.*, 110, 206, 1975.

MEPIRIZOLE

Gas Chromatography

Specimen (mℓ)	S	Column m × mm	Packing (mesh)	Oven temp (°C)	Gas (mℓ/min)	Det.	RT (min)	Internal standard (RT)	Deriv.	Other compounds (RT)	Ref.
B (1) U (50)[a]	1	NA	1.5% OV-17 Chromosorb® W (80/100)	250	Helium[b]	MS-EI	NA	Mepirizole-d_6 (NA)	—	—	1
B (1)	2	1 × 3	1% OV-17 Chromosorb® G (80/100)	160	Nitrogen (60)	ECD	NA	Dichloroanthracene (NA)	Trifluoroacetyl	c	2

[a] Urine extracts were purified by preparative thin-layer chromatography.
[b] Inlet pressure = 1.0 kg/cm^2.
[c] Metabolites of mepirizole have also been chromatographed.

REFERENCES

1. Tanaka, Y. and Sano, M., *Chem. Pharm. Bull.*, 24, 1305, 1976.
2. Tanaka, Y., Esumi, Y., and Sano, M., *Chem. Pharm. Bull.*, 24, 808, 1976.

MEPROBAMATE

Gas Chromatography

Specimen (mℓ)	S	Column m × mm	Packing (mesh)	Oven temp (°C)	Gas (mℓ/min)	Det.	RT (min)	Internal standard (RT)	Deriv.	Other compounds (RT)	Ref.
B (1)	2	1.8 × 2	0.3% PDEAS + 3% DC560 Gas Chrom® p (100/120)	225[a]	Nitrogen (30)	FID	7.5	2, 2-Dipropyl-1, 3-propanedioldicarbamate (11.5)	—	—	1
B, U (1)	1	1.5 × 6	3% SE-30 Chromosorb® W (80/100)	115	Helium (50)	FID	5.5	2-Methyl-2-ethyl-1,3-propanediol (4)	Trimethylsilyl	—	2

[a] The injector is maintained at a lower temperature (185°C) to minimize the decomposition of meprobamate.

REFERENCES

1. Arbin, A., and Ejderfjall, M.-L., *Acta Pharm. Suec.*, 11, 439, 1974.
2. Martis, L. and Levy, R. H., *J. Pharm. Sci.*, 63, 834, 1974.

6-MERCAPTOPURINE

Gas Chromatography

Specimen (mℓ)	S	Column m × mm	Packing (mesh)	Oven temp (°C)	Gas (mℓ/min)	Det.	RT (min)	Internal standard (RT)	Deriv.	Other compounds (RT)	Ref.
B (2)	2	1.5 × NA	10% SE-30 Chromosorb® W (100/120)	135	Nitrogen (100)	FID	21	Theophylline (25)	Methyl[a]	—	1
B (1)	1	1.8 × 4	3% OV-225 Chromosorb® 750 (80/100)	180—200	NA	MS-EI	NA	5-Methyl-N[9]-perdeuteriomethyl-6-mercaptopurine (NA)	Methyl[b]	—	2

High-Pressure Liquid Chromatography

Specimen (mℓ)	S	Column cm × mm	Packing (μm)	Elution	Flow rate (mℓ/min)	Det. (nm)	RT (min)	Internal standard (RT)	Other compounds (RT)	Ref.
Cell extract	3	100 × 1.8	Cation exchange resin M71 (10—12)	E-1	0.13	UV (322)	4.4	—	—	3
Cell extract	3	25 × 4.6	Partisil-10 SAX (NA)	E-2	5	Fl[c] (Ex. 330, Fl. 389)	—	—	6-Thioquanine (6.5) 6-Thioxanthine (8.5) 6-Thioquanosine-monophosphate (15) 6-Thioquanosine-diphosphate (18) 6-Thioquanosine-triphosphate (28)	4

6-MERCAPTOPURINE (continued)

High-Pressure Liquid Chromatography

Specimen (mℓ)	S	Column cm × mm	Packing (μm)	Elution	Flow rate (mℓ/min)	Det. (nm)	RT (min)	Internal standard (RT)	Other compounds (RT)	Ref.
B (0.5)	2	30 × 4	μ-Bondapak® C_{18} (10)	E-3	1.0	UV (313)	4.6	6-Methylthio-2-hydroxypurine (8.4)	6-Thiouric acid (4)	5
B (1)	1	25 × 4.6	LiChrosorb® RP18 (10)	E-4	2.0	UV (325)	5.5	9-Methylmercaptopurine (8)	8-Hydroxymercaptopurine (5) Azathioprine[d] (8)	6

Thin-Layer Chromatography

Specimen (mℓ)	S	Plate (manufacturer)	Layer (mm)	Solvent	Post-separation treatment	Det. (nm)	R_f	Internal standard (R_f)	Other compounds (R_f)	Ref.
D	3	20 × 20 cm Plastic (Merck)	Silica gel 60 F_{254} (0.2)	S-1	Sp: 2 *N* HCl	Visual (UV 366)	0.59	—	6-Thiouric acid (0.03) 8Hydroxy-6-mercaptopurine (0.69) 2-Hydroxy-6-mercaptopurine (0.50) 6-Mercaptopurine riboside (0.58) 6-Mercaptopurine ribosidephosphate (0.08)	7

Note: E-1 = 0.4 *M* Ammonium formate, pH 4.6.

E-2 = i. 5 m*M* Potassium phosphate.
ii. 250 m*M* Potassium phosphate – 500 m*M* potassium chloride.

E-3 = 0.005 *M* 1 Heptanesulfonic acid sodium salt + acetic acid + methanol.
89 : 1 : 10

E-4 = Methanol + acetonitrile + 0.005 *M* phosphate buffer, pH 4 + dithioerythritol (60 mg/ℓ).
1 : 0.5 : 98.5 : 60 mg/ℓ

S-1 = Acetic acid + ethanol
1 : 9

[a] Off-column methylation with trimethylanilinium hydroxide.
[b] Extractive alkylation in the presence of tetrahexylammonium hydroxide as the counter ion.
[c] Thiopurines are oxidized to corresponding sulfonate derivatives with permanganate.
[d] Analysis is carried out under different conditions.

REFERENCES

1. **Bailey, D. G., Wilson, T. W., and Johnson, G. E.,** *J. Chromatogr.*, 111, 305, 1975.
2. **Rosenfeld, J. M., Taguchi, V. Y., Hillcoat, B. L., and Kawai, M.,** *Anal. Chem.*, 49, 725, 1977.
3. **Breter, H. -J. and Zahn, R. K.,** *J. Chromatogr.*, 137, 61, 1977.
4. **Tidd, D. M. and Dedhar, S.,** *J. Chromatogr.*, 145, 237, 1978.
5. **Day, J. L., Tterlikkis, L., Niemann, R., Mobley, A., and Spikes, C.,** *J. Pharm. Sci.*, 67, 1027, 1978.
6. **Ding, T. L. and Benet, L. Z.,** *J. Chromatogr.*, 163, 281, 1979.
7. **Thapliyal, R. C. and Maddocks, J. L.,** *J. Chromatogr.*, 160, 239, 1978.

METAPRAMINE

Gas Chromatography

Specimen (mℓ)	S	Column m × mm	Packing (mesh)	Oven temp (°C)	Gas (mℓ/min)	Det.	RT (min)	Internal standard (RT)	Deriv.	Other compounds (RT)	Ref.
P (1—3) U (1—10)	1	1.5 × 2	3% OV-1 (1) + 3% OV-17 (3) Gas Chrom® Q (100/120)	230	Nitrogen (30)	NPD	4.4	Desipramine (8.7)	a	b	1

[a] Heptafluorbutyryl derivatives are prepared for analysis by electron capture detection.
[b] Retention times of a number of possible metabolites given.

REFERENCE

1. Viala, A. R., Cano, J.-P., Durand, A. G., Erlenmaler, T., and Garreau, R., *Anal. Chem.*, 49, 2354, 1977; *J. Chromatogr.*, 168, 195, 1979.

METFORMIN

Gas Chromatography

Specimen (mℓ)	S	Column m × mm	Packing (mesh)	Oven temp (°C)	Gas (mℓ/min)	Det.	RT (min)	Internal standard (RT)	Deriv.	Other compounds (RT)	Ref.
B (2)	2	2 × 3	3% OV-17 Chromosorb® W (80/100)	250	Nitrogen (40)	NPD	10.5	Propylbiguanide (18)	a	—	1
B (0.1—0.2)	2	1.8 × 4	3% OV-17 Gas Chrom® Q (100/120)	210	Nitrogen (50)	ECD	1.7	Buformin (3.2)	Chlorodifluoroacetyl	—	2

High-Pressure Liquid Chromatography

Specimen (mℓ)	S	Column cm × mm	Packing (μm)	Elution	Flow rate (mℓ/min)	Det. (μm)	RT (min)	Internal standard (RT)	Other compounds (RT)	Ref.
U (2)	2	90 × 2.2	Bondapak® phenyl/corasil (37—50)	E-1	1.0	UV (280)[a]	6	—	—	3

Note: E-1 = Methanol + water.
40 : 60

[a] Conversion to triazene derivatives on treatment with *p*-nitrobenzoyl chloride.

REFERENCES

1. **Borhon, J. and Noel, M.,** *J. Chromatogr.*, 146, 148, 1978.

METFORMIN (continued)

2. Lennard, M. S., Casey, C., Tucker, G. T., and Woods, H. F., *Br. J. Clin. Pharmacol.*, 6, 183, 1978.
3. Ross, M. S. F., *J. Chromatogr.*, 133, 408 1977.

METHADONE

Gas Chromatography

Specimen (mℓ)	S	Column m × mm	Packing (mesh)	Oven temp (°C)	Gas (mℓ/min)	Det.	RT (min)	Internal standard (RT)	Deriv.	Other compounds (RT)	Ref.
B (4) U (0.5—2)	1	1.8 × 2	3% SE-30 Gas Chrom® Q (80/100)	200	Helium (32)	FID	5.4	SKF-525A[a] (10.9)	—	Metabolite 1[b] (3.2) Metabolite 2[c] (2.3)	1
D	3	0.9 × 2	3% SP2250DB Supelcoport (100/120)	235	Helium (35)	FID	1.6	Atropine (2.5)	—	—	2
B (2)	2	1.5 × 1.8	3% DC-560 + 0.3% NPGSe Gas Chrom® P (100/120)	138	Nitrogen (30)	ECD	4	4-(4-Chloro-phenyl)-4-phenyl-2-di-methyl-amino-butane (9.5)	Oxidation[d]	—	3
B (15) Saliva (15) Gastric juice (15)	1	1.8 × 2	1.5% OV-101 Gas Chrom® Q (100-/120)	Temp. Progr.	Nitrogen (30)	FID	5.1	2-Dimethyla-mino-4,4-di-phenyl-5-nona-none (6.4)	—	Metabolite 1[b] (4.2)	4
B (5)	2	1.2 × 2	3% OV-17 Gas Chrom® Q (100/120)	210	Nitrogen (30)	FID	2	SKF-525A[a] (3.6) + *n*-Docosane[e] (0.8)	—	Metabolite 1[b] (1.4) Metabolite 2[c] (1)	5
U (3)	2	1.8 × 2	2% UC W-98 + 1% OV-17 Gas Chrom® Q (100/120)	215	Nitrogen (41)	FID	4	SKF-525A[a] (7.2)	—	Metabolite 1[b] (2.8) Phencyclidine (1.6) Cocaine (5.9)	6
B, U (1)	1	1.8 × 1	10% W-98 Gas Chrom® Q (80/100)	180	Helium (9.2)	MS-CI[f]	7.3	$^{2}H_5$-Methadone (7.3)	—	Metabolite 1[b] (4.3)	7

METHADONE (continued)

High-Pressure Liquid Chromatography

Specimen (mℓ)	S	Column cm × mm	Packing (μm)	Elution	Flow rate (mℓ/min)	Det. (nm)	RT (min)	Internal standard (RT)	Other compounds (RT)	Ref.
D	3	25 × 3.2	LiChrosorb® SI 60 RP 18 (10)	E-1 Gradient	2	UV (280, 254)	13	—	Sodium benzoate[g] (4.8)	8
D	3	30 × 3.9	μ-Bodapak® C_{18} (10)	E-2	1.0	UV (230)	5	Anthracene (8)	—	9

Thin-Layer Chromatography

Specimen (mℓ)	S	Plate (manufacturer)	Layer (mm)	Solvent	Post-separation treatment	Det. (nm)	R_f	Internal standard (R_f)	Other compounds (R_f)	Ref.
U (15)	3	20 × 20 cm (Camag)	Silica gel G (NA)	S-1	Sp: Iodoplatinate reagent	Visual	0.66[h]	—	Metabolite 1[b] (0.74)	10
Tissue	3	5 × 20 cm (Gelman)[i]	Silica gel (1—2)	S-2	Sp: Iodoplatinate reagent	Visual	0.55	—	Methadone-N-oxide (0.23) Metabolite 1[b] (0.36) Metabolite 2[c] (0.97)	11
D	3	20 × 20 cm (Analtech)	Silica gel (0.25)	S-3	E: Iodine vapors	Visual	0.40	—	Methadone-α-naphthalene sulfonate (0.25) Methadone-o-benzoylbenzoate (0.20)	12

Note: E-1 = i. Formic acid + ammonia + water.
10 : 0.25 : 190
ii. Acetonitrile

E-2 = Methanol + water + 1-pentanesulfonic acid, pH 3.5.
75 : 25 : (0.961 g/ℓ)

S-1 = Ethyl acetate + methylene chloride + ammonium hydroxide.
90 : 10 : 0.9

S-2 = i. Ethyl acetate + methanol + ammonium hydroxide (up to 4 cm × 5).
85 : 10 : 5

ii. Benzene + ethyl acetate (up to 17 cm × 1).
95 : 5

iii. Benzene + ethyl acetate + methanol + ammonium hydroxide (up to 17 cm × 1).
80 : 20 : 1.2 : 0.1

S-3 = Isoamyl alcohol + chloroform + acetone + water.
3 : 5 : 1 : 1

[a] β-Diethylaminoethyldiphenylpropyl acetate, obtained from Smith, Kline and French Laboratories.
[b] 2-Ethylidene-1, 5-dimethyl-3-3, diphenylpyrroldine.
[c] 2-Ethyl-5-methyl-3-3, diphenyl-1-pyrroline.
[d] Oxidation to benzophenones with barium peroxide.
[e] External standard added just prior to injection into the gas chromatograph.
[f] Isobutane, 0.5 torr,source pressure.
[g] Preservative in methadone preparation.
[h] Reddish-brown color with iodoplatinate.
[i] Gelman precoated silica gel ITLC type SG sheets were dipped in a slurry of silica gel and $CaSO_4$ to increase the layer thickness.

REFERENCES

1. **Inturrisi, C. E. and Verebely, K.,** *J. Chromatogr.*, 65, 361, 1972; *Res. Commun. Chem. Pathol. Pharmacol.*, 6, 353, 1973.
2. **Choulis, N. H. and Papadopoulos, H.,** *J. Chromatogr.*, 106, 180, 1975.
3. **Hartvig, P. and Näslund, B.,** *J. Chromatogr.*, 111, 347, 1975.
4. **Lynn, R. K., Leger, R. M., Gordon, W. P., Olsen, G. D., and Gerber, N.,** *J. Chromatogr.*, 131, 329, 1977.

METHADONE (continued)

5. Thompson, B. C. and Caplan, Y. H., *J. Anal. Toxicol.*, 1, 66, 1977.
6. Jain, N. C., Chinn, D. M., Sneath, T. C., and Budd, R. D., J. Anal. Toxicol., *1, 192, 1977.*
7. Hachey, D. L., Kreek, M. J., and Mattson, D. H., *J. Pharm. Sci.*, 66, 1579, 1977.
8. Beasley, Sr., T. H., and Ziegler, H. W., *J. Pharm. Sci.*, 66, 1749, 1977.
9. Hsieh, J.-W., Ma, J. K. H., O'Donnel, J. P., and Choulis, N. H., *J. Chromatogr.*, 161, 366, 1978.
10. Jain, N. C., Leung, W. J., Budd, R. D., and Sneath, T. C., *J. Chromatogr.*, 103, 85, 1975.
11. Davis, C. M. and Fenimore, D. C., *J. Chromatogr.*, 104, 193, 1975.
12. Choulis, N. H., *J. Chromatogr.*, 124, 172, 1976.

METHANESULFOANILIDE

High-Pressure Liquid Chromatography

Specimen (mℓ)	S	Column cm × mm	Packing (μm)	Elution	Flow rate (mℓ/min)	Det. (nm)	RT (min)	Internal standard (RT)	Other compounds (RT)	Ref.
B (2)	1	30 × 6.4	μ-Bondapak® C_{18} (NA)	E-1	2.7	UV (254)	4.8	2-(4′-Chlorophenoxy)-4-nitromethanesulfoanilide (6)	—	1

Note: E-1 = Acetonitrile + water.
50 : 50

REFERENCE

1. **Chang, S. F., Miller, A. M., and Ober, R. E.,** *J. Pharm. Sci.*, 66, 1700, 1977.

METHAPYRILENE

Gas Chromatography

Specimen (mℓ)	S	Column m × mm	Packing (mesh)	Oven temp (°C)	Gas (mℓ/min)	Det.	RT (min)	Internal standard (RT)	Deriv.	Other compounds (RT)	Ref.
B (1)	2	1.2 × 3	1% HI-EFF-8BP + 10% SE-52 Gas chrom® Q (80/100)	220	Helium (NA)	FID	3.5	Phenothiazine (6.8)	—	—	1

REFERENCE

1. Schirmer, R. E. and Pierson, R. J., *J. Pharm. Sci.*, 62, 2052, 1973.

METHAQUALONE

Gas Chromatography

Specimen (mℓ)	S	Column m × mm	Packing (mesh)	Oven temp (°C)	Gas (mℓ/min)	Det.	RT (min)	Internal standard (RT)	Deriv.	Other compounds (RT)	Ref.
B (2)	2	1.5 × 2	3% OV-1 Gas Chrom® Q (100/120)	210	Helium (30)	FID	2.5	*N*-Propionyl-2-benzoyl-4-chloroaniline (4.2)	—	—	1
B, U (20)	3	3 × 2	3% OV-225 Chromosorb® (80/100)	Temp. Progr.	NA	MS-EI	NA	—	Trimethyl-silyl	1-Hydroxymetha-qualone (21.5) 10-Hydroxymetha-qualone (22) 2-Hydroxymetha-qualone (30) 3-Hydroxymetha-qualone (31)	2
U (5)	3	50 × 0.76[a]	SE-30	205	NA	MS-EI	NA	—	—	1-Hydroxymetha-qualone[b] (46) 10-Hydroxymetha-qualone (48) 2-Hydroxymetha-qualone (55) 3-Hydroxymetha-qualone (56)	3
U (0.5)	2	1.8 × 2	3% SE-30 Gas Chrom® Q (80/100)	255	Nitrogen (35)	FID[d]	2.5	SKF-525-A[c] (5)	Heptafluo-robutyryl	e	4

METHAQUALONE (continued)

Thin-Layer Chromatography

Specimen (mℓ)	S	Plate (manufacturer)	Layer (mm)	Solvent	Post-separation treatment	Det. (nm)	R_f	Internal standard (R_f)	Other compounds (R_f)	Ref.
U (5)	3	20 × 20 cm (Merck)	Silica gel F_{254} (NA)	S-1	SP: Iodoplatinate reagent	Visual	0.64	—	f	5

Note: S-1 = Diethyl ether.

[a] Support coated open tubular column.
[b] Nomenclature of metabolites as in Reference 2.
[c] β-Diethylaminoethyl diphenyl propyl acetate.
[d] A nitrogen specific detector was also used.
[e] Retention times of various hydroxymetabolites given.
[f] R_f values and alternative solvent system for the separation of four major hydroxy metabolites are given.

REFERENCES

1. Evenson, M. A. and Lensmeyer, G. L., *Clin. Chem.*, 20, 249, 1974.
2. Bonnichsen, R., Dimberg, R., Marde, Y. and Ryhage, R., *Clin. Chim. Acta*, 60, 67, 1975.
3. Kazyak, L., Kelley, J. A., Cella, J. A., Droege, R. E., Hilpert, L. R., and Permisohn, R. C., *Clin. Chem.*, 23, 2001, 1977.
4. Mulé, S. J., Kogan, M., and Jukofsky, D., *Clin. Chem.*, 24, 1473, 1978.
5. Sleeman, H. K., Cella, J. A., Harvey, J. L., and Beach, D. J., *Clin. Chem.*, 21, 76, 1975.

METHIMAZOLE

Gas Chromatography

Specimen (mℓ)	S	Column m × mm	Packing (mesh)	Oven temp (°C)	Gas (mℓ/min)	Det.	RT (min)	Internal standard (RT)	Deriv.	Other compounds (RT)	Ref.
B (1)	1	2 × 4	10% Apiezon® L + 5% KOH Chromosorb® W (100/120)	195	Helium (40)	NPD	3	6-Hydroxypyridazin-3 (2H-one) (3.8)	Methyl[a]	—	1

High-Pressure Liquid Chromatography

Specimen (mℓ)	S	Column cm× mm	Packing (μm)	Elution	Flow rate (mℓ/min)	Det. (nm)	RT (min)	Internal standard (RT)	Other compounds (RT)	Ref.
B (1)	2	100 × 4.6	Spherisorb Si[b] (10)	E-1	1.2	UV (254)	3.5	Benzamide (5.5)	—	2

Note: E-1 = Chloroform + methanol.
99 : 1

[a] Tetramethylammonium hydroxide.
[b] The top 1.5 cm of the column is packed with alumina (10 μm).

REFERENCES

1. **Bending, M. R. and Stevenson, D.,** *J. Chromatogr.*, 154, 267, 1978.
2. **Skellern, G. G., Knight, B.B., and Stenlake, J. B.,** *J. Chromatogr.*, 124, 405, 1976.

METHINDIONE

Gas Chromatography

Specimen (mℓ)	S	Column m × mm	Packing (mesh)	Oven temp (°C)	Gas (mℓ/min)	Det.	RT (min)	Internal standard (RT)	Deriv.	Other compounds (RT)	Ref.
B (2)	1	NA	3% OV-17 Gas Chrom® P (100/120)	160	Nitrogen (30)	ECD	5	2-Propyl analog (6.5)	—	—	1

REFERENCE

1. Vessman, J., Strömberg, S., and Freij, G., *J. Chromatogr.*, 94, 239, 1974.

METHOTREXATE

High-Pressure Liquid Chromatography

Specimen (mℓ)	S	Column cm × mm	Packing (μm)	Elution	Flow rate (mℓ/min)	Det. (nm)	RT (min)	Internal standard (RT)	Other compounds (RT)	Ref.
D	3	50 × 2.1	Vydac® (30—44)	E-1	0.6	UV (254)	9	—	N^{10}-Methylfolic acid (5)	1
D	3	25 × 9	DE 52 DEAE Cellulose (NA)	E-2 gradient	1.7	UV (254)	108	—	a	2
B (1)	1	25 × 4.6	Partisil®-10-SAX (10)	E-3	1.2	UV (315)	7.3	b (8.9)	7-Hydroxymethotrexate (11.3)	3
D U (0.01)	2	30 × 4	μ-Bondapak® C_{18} (10)	E-4	0.45	UV (254)	35	—	7-Hydroxymethotrexate (48)	4
B (NA)	2	25 × 4.6[c]	Partisil®-10-SAX (10)	E-5	2.5	UV (306)	6	—	—	5

Note: E-1 = $NaClO_4$ (0.3 mole) + NaH_2PO_4 (0.15 mole) + Na_2HPO_4 (0.15 mole) + acetonitrile (15 mℓ/ℓ.)

E-2 = i. 0.1 *M* Ammonium bicarbonate.
ii. 0.4 *M* Ammonium bicarbonate. Linear gradient: 0.5%/min.

E-3 = 0.025 *M* Sodium phosphate buffer, pH 7.

E-4 = 5 m*M* Tetrabutylammonium phosphate, pH 7.5.

E-5 = 0.05 *M* Phosphate buffer containing 0.01 *M* sodium chloride + methanol.
4 : 1

[a] A number of impurities in parenteral methotrexate dosage forms have been separated and identified.
[b] *N*-(4[(2, 4-Diamino-6-quinzolinyl) methylamino] benzoyl) aspartic acid.
[c] A precolumn (4.6 cm × 3 mm packed with 10 *M* RP-8) was used for the concentration of methotrexate from 1 mℓ of deproteinized plasma injection.

METHOTREXATE (continued)

REFERENCES

1. Chatterji, D. C. and Gallelli, J. F., *J. Pharm. Sci.*, 66, 1219, 1977.
2. Hignite, C. E., Shen, D. D., and Azarnoff, D. L., *Cancer Treat. Rep.*, 62, 13, 1978.
3. Watson, E., Cohen, J. L., and Chan, K. K., *Cancer Treat. Rep.*, 62, 381, 1978.
4. Wisnicki, J. L., Tong, W. P., and Ludlum, D. B., *Cancer Treat. Rep.*, 63, 529, 1978.
5. Lankelma, J. and Poppe, H., *J. Chromatogr.*, 149, 587, 1978.

METHOTRIMEPRAZINE

Gas Chromatography

Specimen (mℓ)	S	Column m × mm	Packing (mesh)	Oven temp (°C)	Gas (mℓ/min)	Det.	RT (min)	Internal standard (RT)	Deriv.	Other compounds (RT)	Ref.
B (1—6)	2	1.8 × 2	3% OV-17 Supelcoport (80/100)	255	Nitrogen (25)	FID	3.5	Trifluoperazine (4.5)	—	Methotrimeprazine sulfoxide (14.5)	1

REFERENCE

1. Dahl, S. G. and Jacobsen, S., *J. Pharm. Sci.*, 65, 1329, 1976.

METHOXYPHENAMINE

Gas Chromatography

Specimen (mℓ)	S	Column m × mm	Packing (mesh)	Oven temp (°C)	Gas (mℓ/min)	Det.	RT (min)	Internal standard (RT)	Deriv.	Other compounds (RT)	Ref.
U (100—200)	3	1.8 × 2	5% OV-25 Chromosorb® W (80/100)	180	Nitrogen (70)	FID	2.5	—	Trifluoroacetyl	*N*-Demethylated methoxyphenamine (2) *O*-Demethylated methoxyphenamine (4)	1
U (NA)	3	NA	3% SE-30 Anakhrom (NA)	140	NA	FID	9	—	—	*N*-Demethylated methoxyphenamine (7.3) *O*-Demethylated methoxypehnamine (12.5)	2

REFERENCES

1. Midha, K. K., Cooper, J. K., McGilveray, I. J., Coutts, R. T., and Darve, R., *Drug. Metab. Dispos.*, 4, 568, 1976.
2. Chundela, B. and Slechtova, R., *J. Chromatogr.*, 119, 609, 1976; 144, 284, 1977.

8-METHOXYPSORALEN

Gas Chromatography

Specimen (mℓ)	S	Column m × mm	Packing (mesh)	Oven temp (°C)	Gas (mℓ/min)	Det.	RT (min)	Internal standard (RT)	Deriv.	Other compounds (RT)	Ref.
B (2)	1	0.9 × 2	3% OV-225 Gas Chrom® Q (100/120)	195	Nitrogen (20)	ECD	8	8-Butoxy psoralen (15.5)	—	—	1
B (0.5)	1	2 × 2	3% OV-17 Chromosorb® 750 (80/100)	220	Argon 95 + Methane 5 (NA)	ECD	4.5	5,8-Dimethoxypsoralen (7)	—	—	2

High-Pressure Liquid Chromatography

Specimen (mℓ)	S	Column cm × mm	Packing (μm)	Elution	Flow rate (mℓ/min)	Det. (nm)	RT (min)	Internal standard (RT)	Other compounds (RT)	Ref.
B (1)	1	25 × 4.6	Partisil® (Silica) (10)	E-1	2.2	UV (254)	3.8	—	—	3

Thin-Layer Chromatogaphy

Specimen (mℓ)	S	Plate (manufacturer)	Layer (mm)	Solvent	Post-separation treatment	Det. (nm)	R_f	Internal standard (R_f)	Other compounds (R_f)	Ref.
B(1)	2	20 × 20 cm (Merck)	Silica gel (0.25)	S-1	—	Fl: Reflectance (Ex.) a, Fl: b	0.6	—	—	4
B (5)	1	20 × 20 cm (Analtech)	Silica gel G (0.25)	S-2	—	Fl: Reflectance (Ex: 310, Fl: 540)	NA	—	—	5

8-METHOXYPSORALEN (continued)

Note: E-1 = Acetonitrile + methylene chloride.
5 : 95

S-1 = Chloroform + methanol.
97 : 3

S-2 = Benzene + ethyl acetate.
9 : 1

[a] Primary filter: UVB (transmission of all mercury lines between 240 and 400 nm).
[b] Secondary filter: U3 (transmission > 500 nm).

REFERENCES

1. Ehrsson, H., Eksborg, S., Wallin, I., Kallberg, N., and Swanbeck, G., *J. Chromatogr.*, 140, 157, 1977.
2. Schmid, J. and Koss, F. W., *J. Chromatogr.*, 146, 498, 1978.
3. Puglisi, C. V., de Silva, J. A. F., and Meyer, J. C., *Anal. Lett.*, 10, 39, 1977.
4. Herfst, M. J., Koot-Gronsveld, E. A. M., and de Wolff, F. A., *Arch. Dermatol. Res.*, 262, 1, 1978.
5. Chakarbarti, S. G., Gooray, D. A., and Kenny, J. A., Jr., *Clin. Chem.* 24, 1155, 1978.

METHYLDOPA

Gas Chromatography

Specimen (mℓ)	S	Column m × mm	Packing (mesh)	Oven temp (°C)	Gas (mℓ/min)	Det.	RT (min)	Internal standard (RT)	Deriv.	Other compounds (RT)	Ref.
D	3	1.8 × NA	5% OV-101 Chromosorb® W (100/120)	170	Nitrogen (60)	FID	15.7	Dibenzylsuccinate (22)	Trimethylsilyl	Methylphenylalanine (2.2) α-Methyl-3-methoxy phenylalanine (4.6) α-Methyl-3-hydroxy phenylalanine (6.8) α-Methyl-4-hydroxy phenylalanine (7.7)	1

High-Pressure Liquid Chromatography

Specimen (mℓ)	S	Column cm × mm	Packing (μm)	Elution	Flow rate (mℓ/min)	Det. (nm)	RT (min)	Internal standard (RT)	Other compounds (RT)	Ref.
U (30)	2	30 × 4	μ-Bondapak® C_{18} (10)	E-1	2.0	UV (280)	5.2	3,4-Dihydroxybenzylamine (2.7)	Norepinephrine (2.3) Epinephrine (2.8) L-Dopa (3.3) Dopamine (3.7)	2
B (0.2)	2	30 × 4	μ-Bondapak® C_{18} (10)	E-2	2.0	Electrochemical	9.2	Norepinephrine (2.1)	L-Dopa (4.1) Dopamine (5.5) α-Methyldopamine (13.2)	3
B (1)	2	50 × 2	Vydac® SEX (30—44)	E-3	0.4	Electrochemical	2	—	α-Methyldopamine (19)	4

METHYLDOPA (continued)

Note: E-1 = 0.17 *M* Acetic acid + 0.2 *M* sodium acetate.
92 : 8

E-2 = 0.1 *M* Nitric acid, pH 2.3 (pH adjusted with 1 *N* NaOH).

E-3 = 20 m*M* Ammonium dihydrogen phosphate (+ 0.1 m*M* EDTA), pH 2.5.

REFERENCES

1. **Watson, J. R. and Lawrence, R. C.,** *J. Chromatogr.,* 103, 63, 1975.
2. **Mell, L. D. and Gustafson, A. B.,** *Clin. Chem.,* 24, 23, 1978.
3. **Freed, C. R. and Asmus, P. A.,** *J. Neurochem.,* 32, 163, 1979.
4. **Cooper, M. J., O'Dea, R. F., and Mirkin, B. L.,** *J. Chromatogr.,* 162, 601, 1979.

METHYLGLUCAMINE

High-Pressure Liquid Chromatography

Specimen (mℓ)	S	Column cm × mm	Packing (μm)	Elution	Flow rate (mℓ/min)	Det. (nm)	RT (min)	Internal standard (RT)	Other compounds (RT)	Ref.
U (NA)	3	100 × 2[a]	Cation exchange resin (NA)	E-1	0.83	Refractive index	10	—	—	1

Note: E-1 = 0.005 *M* KH_2PO_4, pH 7.2.

[a] Column temperature = 45°C.

REFERENCE

1. Nahlovsky, B. D. and Lang. J. H., *J. Chromatogr.*, 101, 225, 1974.

METHYLOXAZEPAM

Gas Chromatography

Specimen (mℓ)	S	Column m × mm	Packing (mesh)	Oven temp (°C)	Gas (mℓ/min)	Det.	RT (min)	Internal standard (RT)	Deriv.	Other compounds (RT)	Ref.
B (0.1—0.2)	1	2 × 4	3% OV-11 Chromosorb® Q (100/200)	260	Nitrogen (50)	ECD	1.9	Diazepam (NA)	Trimethyl-silyl	—	1

REFERENCE

1. Belvedere, G., Tognoni, G., Frigerio, A., and Morselli, P. L., *Anal. Lett.*, 5, 531, 1972.

METHYLPENTYNOL CARBAMATE

Gas Chromatography

Specimen (mℓ)	S	Column m × mm	Packing (mesh)	Oven temp (°C)	Gas (mℓ/min)	Det.	RT (min)	Internal standard (RT)	Deriv.	Other compounds (RT)	Ref.
B (2)	2	1.5 × 4	4% CDMS Diatomite CLQ (100/120)	180	Nitrogen (60)	FID	3	3,5 Xylenol (6)	—	3-Methylpentyne-3,4-diol[a]	1

[a] Conditions for the analysis of this metabolite are described; using the same column, retention time at 140°C = 2.6 min.

REFERENCE

1. Grove, J. and Martin, B. K., *J. Chromatogr.*, 133, 267, 1977.

METHYLPHENIDATE

Gas Chromatography

Specimen (mℓ)	S	Column m × mm	Packing (mesh)	Oven temp (°C)	Gas (mℓ/min)	Det.	RT (min)	Internal standard (RT)	Deriv.	Other compounds (RT)	Ref.
(2)	2	1.8 × 2	3% OV-17 Gas Chrom® Q (100/120)	170	Nitrogen (25—30)	FID	4	Diphenhydramine (7)	Methyl	Ritalinic acid (4)[a]	1
B (4)	3	1.2 × 4	3% OV-1 Chromosorb® G (80/100)	190	Helium (40)	ECD	1.3	—	i. Reduction[b] ii. Pentafluoropropionyl	—	2
U (10)	3	3 × 3	2% OV-17 Gas Chrom® Q (100/120)	160	Nitrogen (40)	ECD	32.6	—	Pentafluoropropionyl	c	3
B (2) Tissue	1	0.9 × 2	3% OV-17 Gas Chrom® Q (NA)	170	Helium (35)	MS-EI	2.8	Ethylphenidate (3.4)	Trifluoroacetyl	—	4
B (2)	2	1.8 × 2	3% OV-1 Gas Chrom® Q (100/120)	180	Helium (30)	NPD	3.7	Ethyphenidate (4.4)	Trifluoroacetyl	Ritalinic acid (3.7)[a]	5
U (0.5)	3	2 × 2	2.5% OV-225 Chromosorb® W (80/100)	180	Nitrogen (30)	ECD	—	—	Pentafluorobenzyl[d]	Ritalinic acid (5.3)	6

High-Pressure Liquid Chromatography

Specimen (mℓ)	S	Column cm × mm	Packing (μm)	Elution	Flow rate (mℓ/min)	Det. (nm)	RT (min)	Internal standard (RT)	Other compounds (RT)	Ref.
B (0.3)	1	30 × 4	μ-Bondapak® C_{18}[e] (10)	E-1	2.0	UV (192)	—	α, α-Dimethyl-β-methylsucci-	Ritalinic acid (14)	7

B (0.3)	1	30 × 4	μ-Bondapak® C_{18}[e] (10)	E-2	1.6	UV (192)	6	nimide (10) 4, 5-Diphenylimidazole (17)	—	8

Thin-Layer Chromatography

Specimen (mℓ)	S	Plate (manufacturer)	Layer (mm)	Solvent	Post-separation treatment	Det. (nm)	R_f	Internal standard (R_f)	Other Compounds (R_f)	Ref.
U (1)	3	NA (Machery-Nagel)	Polygram Sil G/254 (NA)	S-1	i. Sp: 0.2 g Ninhydrin in 100 mℓ of acetone ii. Heat at 120°C for 15 min.	Visual	0.32[f]	Ephedrine (0.37)	—	6

Note: E-1 = 20 m*M* Potassium phosphate buffer, pH 3.8 + acetonitrile.
93 : 7

E-2 = 20 m*M* Potassium phosphate buffer, pH 3.5 + acetonitrile.
80 : 20

S-1 = i. Acetone + acetic acid (×2).
98 : 2
ii. Methanol + chloroform + acetic acid.
50 : 50 : 2

[a] Ritalinic acid is converted to methylphenidate by treatment with diazomethane.
[b] With $LiAlH_4$.
[c] Retention times of pentafluoropropionyl derivatives of some secondary amines using different columns at different temperatures are given.
[d] Extractive alkylation in the presence of pentafluorobenzyl bromide and tetrabutylammonium hydrogen sulfate.
[e] Column temperature = 40°C.
[f] Violet color.

METHYLPHENIDATE (continued)

REFERENCES

1. Wells, R., Hammond, K. B., and Rodgerson, D. O., *Clin. Chem.*, 20, 440, 1974.
2. Huffman, R., Blake, J. W., Ray, R., Noonan, J., and Murdick, P. W., *J. Chromatogr. Sci.*, 12, 382, 1974.
3. Delbeke, F. T. and Debackere, M., *J. Chromatogr.*, 106, 412, 1975.
4. Gal, J., Hodshon, B. J., Pintauro, C., Flamm, B. L., and Cho, A. K., *J. Pharm. Sci.*, 66, 866, 1977.
5. Hungund, B. L., Hanna, M., and Winsberg, B. G., *Commu. Psychopharmacol.*, 2, 203, 1978.
6. van Boven, M. and Dainens, P., *J. Forens. Sci.*, 24, 55, 1979.
7. Soldin, S. J., Hill, B. M., Chan, Y.-P. M., Swanson, J. M., and Hill, G. J., *Clin. Chem.*, 25, 51, 1979.
8. Soldin, S. J., Chan, Y.-P. M., Hill, B. M., and Swanson, J. M., *Clin. Chem.*, 25, 401, 1979.

METHYSALICYLATE

Gas Chromatography

Specimen (mℓ)	S	Column m × mm	Packing (mesh)	Oven temp (°C)	Gas (mℓ/min)	Det.	RT (min)	Internal standard (RT)	Deriv.	Other compounds (RT)	Ref.
D	3	1.8 × 3	20% Carbowax 20 *M* Gas Chrom® Q (80/100)	155	Helium (55)	FID	3.1	*n*-Nonadecane (5)	—	Menthol (1.8)	1

REFERENCE

1. Sapio, J. P., Sethachutkul, K., and Moody, J. E., *J. Pharm. Sci.*, 68, 506, 1979.

METHYPRYLON

Gas Chromatography

Specimen (mℓ)	S	Column m × mm	Packing (mesh)	Oven temp (°C)	Gas (mℓ/min)	Det.	RT (min)	Internal standard (RT)	Deriv.	Other compounds (RT)	Ref.
B (0.5) U (0.5)[a]	3	1.8 × 2	3% OV-17 Gas Chrom® Q (80/100)	Temp. Progr.	Nitrogen (10)	FID	7.3	1, 3-Dibutyl-5,5-diethylbarbituric acid (9.7)	—	Methyprylon metabolite[b] (7.7)	1
B, U (2)	2	50 × 0.5[c]	Superox 4[d]	200	Hydrogen (5)	FID	6	Dihydroprylone (6.5)	—	Methyprylon metabolite[b] (8.5)[e]	2

[a] Urine is diluted 1:100.
[b] 2,4-Dioxo-3,3-diethyl-5-methyl-1,2,3,4-tetrahydropyridine.
[c] Wall-coated open tubular column.
[d] Equivalent to Carbowax® 20 *M.*
[e] Other metabolites of methyprylon have also been separated.

REFERENCES

1. Bridges, R. R. and Peat, M. A., *J. Anal. Toxicol.*, 3, 21, 1979.
2. van Boven, M. and Sunshine, I., *J. Anal. Toxicol.*, 3, 174, 1979.

METOCLOPRAMIDE

High-Pressure Liquid Chromatography

Specimen (mℓ)	S	Column cm × mm	Packing (μm)	Elution	Flow rate (mℓ/min)	Det. (nm)	RT (min)	Internal standard (RT)	Other compounds (RT)	Ref.
B	1	15 × 0.5	Silica gel M 131 (NA)	E-1	2.0	UV (280)	2.8	4-Amino-5-chloro-*N*-[2-(propylamino)-ethyl]-2-methoxy benzamide (4.5)	Monodeethylated metoclopramide (7.5) Dideethylated metoclopramide (5.7)	1

Thin-Layer Chromatography

Specimen (mℓ)	S	Plate (manufactrer)	Layer (mm)	Solvent	Post-separation treatment	Det. (nm)	R_f	Internal standard (R_f)	Other compounds (R_f)	Ref.
Tissue	1	5 × 20 20 × 20 (Merck)	Silica gel 60 F_{254} (0.25)	S-1[a]	Sp: i. 1% Sodium nitrite in 5 *N* HCl ii. 0.4% *N*(-1-Naphthyl) ethylenediammonium chloride in methanol	Visible Reflectance (550)	0.46	—	Monode-ethylated[b] metoclopramide (0.19) Dide-ethylated metoclopramide (0.11)	2

Note: E-1 = Methanol + chloroform + ammonium hydroxide.
30 : 70 : 0.5

S-1 = 1,2-Dichloroethane + ethanol + ammonia.
70 : 15 : 2

[a] Alternative solvent systems have been described.
[b] Rf values of other metabolites also given.

METOCLOPRAMIDE (continued)

REFERENCES

1. Teng, L., Bruce, R. B., and Dunning, L. K., *J. Pharm. Sci.*, 66, 1615, 1977.
2. Huizing, G., Beckett, A. H., and Segura, J., *J. Chromatogr.*, 172, 227, 1979.

METOLAZONE

High-Pressure Liquid Chromatograhy

Specimen (mℓ)	S	Column cm × mm	Packing (μm)	Elution	Flow rate (mℓ/min)	Det. (nm)	RT (min)	Internal standard (RT)	Other compounds (RT)	Ref.
U (5)	3	183 × 3	Durapak (NA)	E-1	1	UV (254)	NA	—	—	1

Note: E-1 = Chloroform + isopropyl alcohol.
95 : 5

REFERENCE

1. **Hinsvark, O. N., Zazulak, W., and Cohen, A. I.,** *J. Chromatogr. Sci.*, 10, 379, 1972.

METOPRINE

High-Pressure Liquid Chromatography

Specimen (mℓ)	S	Column cm × mm	Packing (μm)	Elution	Flow rate (mℓ/min)	Det. (nm)	RT (min)	Internal standard (RT)	Other compounds (RT)	Ref.
Tissue B (0.1)	2	30 × 4	μ-Bondapak® C_{18} (10)	E-1	1.5	UV (280)	5.5	Etoprine (7)	—	1

Note: E-1 = Methanol + 0.02 *M* phosphate buffer, pH 7.5.
65 : 35

REFERENCE

1. Levin, E. M., Meyer, R. B., Jr., and Levin, V. A., *J. Chromatogr.*, 156, 181, 1978.

METRONIDAZOLE

Gas Chromatography

Specimen (mℓ)	S	Column m × mm	Packing (mesh)	Oven temp (°C)	Gas (mℓ/min)	Det.	RT (min)	Internal standard (RT)	Deriv.	Other compounds (RT)	Ref.
B (2)	2	1.8 × 4	3% OV-1 Gas Chrom® Q (NA)	160	Nitrogen (70)	FID	4.1	Myristyl alcohol (8.1)	Trimethylsilyl	2-Methyl-5-nitroimidazole-1-ylacetic acid (5.1)	1

High-Pressure Liquid Chromatography

Specimen (mℓ)	S	Column cm × mm	Packing (μm)	Elution	Flow rate (mℓ/min)	Det. (nm)	RT (min)	Internal standard (RT)	Other compounds (RT)	Ref.
B (NA)	1	30 × 3.9	μ-Bondapak® C_{18} (10)	E-1	2.3	UV (313)	5.7	—	2-Methyl-5-nitroimidazole-1-ylacetic acid (4.5) 1-(2-Hydroxyethyl)-2-hydroxymethyl-5-nitroimidazole (3.3)	2
B, U (NA)	1	30 × 4	μ-Bondapak® C_{18} (10)	E-2	2.0	UV (324)	5	—	Misonidazole (5) Desmethylmisonidazole (2)	3
B (0.1)	1	15 × 4.6	Spherisorb 5 ODS (5)	E-3	1.0	UV (320)	5	1-(3-Hydroxypropyl)-2-methyl-5-nitroimidazole (7.2)	Tinidazole (9)	4
B (0.1)	1	30 × 4	μ-Bondapak® C_{18} (10)	E-4	1.5	UV (320)	6.4	—	Tinidazole (9.2) Desmethylmisonidazole	5

METRONIDAZOLE (continued)

Thin-Layer Chromatography

Specimen (mℓ)	S	Plate (manufacturer)	Layer (mm)	Solvent	Post-separation treatment	Det. (nm)	R_f	Internal standard (R_f)	Other compounds (R_f)	Ref.
B (2)	1	20 × 20 cm (Merck)	Silica gel F_{254} (0.25)	S-1	—	Quenching of fluorescence (254)	0.15	—	Tinidazole (0.35)	6

Note: E-1 = Methanol + acetonitrile + 0.005 *M* KH_2PO_4, pH 4.
4 : 3 : 93

E-2 = Acetonitrile + 0.01 m*M* phosphate buffer, pH 4.
8 : 92

E-3 = Acetonitrile + 0.01 *M* phosphate buffer, pH 5.5.
15 : 85

E-4 = Acetonitrile + 20 m*M* acetate buffer, pH 4.0.
7 : 93

S-1 = Chloroform + acetic acid.
9 : 1

REFERENCES

1. Wood, N. F., *J. Pharm Sci.*, 64, 1048, 1975.
2. Wheeler, L. A., de Meo, M., Halula, M., George, L., and Heseltine, P., *Antimicrob. Agents Chemother.*, 13, 205, 1978.
3. Marques, R. A., Stafford, B., Flynn, N., and Sadee, W., *J. Chromatogr.* 146, 163, 1978.
4. Lanbeck, K. and Lindström, B., *J. Chromatogr.*, 162, 117, 1979.
5. Hackett, L. P. and Dusci, L. J., *J. Chromatogr.*, 175, 347, 1979.
6. Welling, P. G. and Monro A. M., *Arzneim. Forsch.*, 22, 2128, 1972.

MEXITELINE

Gas Chromatography

Specimen (mℓ)	S	Column m × mm	Packing (mesh)	Oven temp (°C)	Gas (mℓ/min)	Det.	RT (min)	Internal standard (RT)	Deriv.	Other compounds (RT)	Ref.
B (1)	1	3 × 4	1% HI EFF 8BP Gas Chrom® Q (100/120)	180	Argon 95 + Methane 5 (45)	ECD	3.3	Alprenolol (5.2)	Heptafluorobutyryl	—	1
B (1)	1	1.5 × 4	2.5% OV-101 Gas Chrom® Q (70/80)	150	Nitrogen (38)	NPD	2	*p*-Nitroanisole (3)	—	—	2
B (0.1)	1	4 × 4	2% OV-17 + 1% OV-275 Chromosorb® G (80/100)	180	Nitrogen (40)	ECD	11.7	K0768[a] (12.9)	Heptafluorobutyryl	3	
B (0.2)	2	2.1 × 4	10% Apiezon L + 2% KOH Chromosorb® W (80/100)	195	Nitrogen (70)	FID	4.3	2,7-Dimethylquinoline (7)	—	Nicotine[b] (3.8)	4

[a] 1-(2,4-Dimethylphenoxy)-2-aminopropane.
[b] Conditions for the assay of lignocaine are also described.

REFERENCES

1. Willox, S. and Singh, B. N., *J. Chromatogr.*, 128, 196, 1976.
2. Bradbrook, I. D., James, C. and Rogers, H. J., *Br. J. Clin. Pharmacol.*, 4, 380, 1977.
3. Frydman, A., Lafarge, J.-P., Vial, F., Rulliere, R., and Alexander, J.-M., *J. Chromatogr.*, 145, 401, 1978.
4. Holt, D. W., Flanagen, R. J., Hayler, A. M., and Loizou, M., *J. Chromatogr.*, 169, 295, 1979.

MIANSERIN

Gas Chromatography

Specimen (mℓ)	S	Column m × mm	Packing (mesh)	Oven temp (°C)	Gas (mℓ/min)	Det.	RT (min)	Internal standard (RT)	Deriv.	Other compounds (RT)	Ref.
B (1—2)	1	4 × 2	1% JXR Gas Chrom® Q (80/100)	260	Helium (30)	MS-EI[a]	2.5	[$^2H_2$]-Mianserin (2.5)	—	—	1

[a] The plasma extracts were purified by preparative liquid chromatography, using a 30 cm × 4 mm μ-Porasil® column (10 μm). The column was eluted with *n*-hexane-isopropanol (80:20) containing 4% ethanol and 0.1% ammonia at a flow rate of 2 mℓ/min. The eluate was monitored at 280 nm.

REFERENCE

1. **de Ridder, J. J., Koppens, P. C. J. M., and van Hal, H. J. M.,** *J. Chromatogr.*, 143, 289, 1977.

MICONAZOLE

Gas Chromatography

Specimen (mℓ)	S	Column m × mm	Packing (mesh)	Oven temp (°C)	Gas (mℓ/min)	Det.	RT (min)	Internal standard (RT)	Deriv.	Other compounds (RT)	Ref.
Spin	2	1.8 × 3	3% OV-17 Gas Chrom® Q (100/120)	250	Argon (NA) + Methane (NA) (9)	ECD	3.2	8-Chlorotheophylline (1.2)	Methyl[a]	Clotrimazole[b]	1

[a] On-column methylation with tetramethylammonium hydroxide.
[b] Conditions for the separate analysis of clotrimazole are also described.

REFERENCE

1. **Wallace, S. M., Shah, V. P., Riegelman, S., and Epstein, W. L.,** *Anal. Lett.*, B11, 461, 1978.

MISONIDAZOLE

High-Pressure Liquid Chromatography

Specimen (mℓ)	S	Column cm × mm	Packing (μm)	Elution	Flow rate (mℓ/min)	Det. (nm)	RT (min)	Internal standard (RT)	Other compounds (RT)	Ref.
B (NA)	1	30 × 3.9	μ-Bondapak® C_{18} (10)	E-1	2	UV (313)	4	RO-07-0913[a] (7.3)	O-Desemethylmisonidazole (2.2)	1

Note: E-1 = Methanol + water.
19 : 81

[a] 1-(2-Nitroimidazol-1-yl)-3-ethoxypropan-2-ol.

REFERENCE

1. **Workman, P., Little, C. J., Marten, T. R., Dale, A. D., and Ruane, R. J.,** *J. Cromatogr.*, 145, 507, 1978.

MOLSIDOMINE

High-Pressure Liquid Chromatography

Specimen (mℓ)	S	Column cm × mm	Packing (μm)	Elution	Flow rate (mℓ/min)	Det. (nm)	RT (min)	Internal standard (RT)	Other compounds (RT)	Ref.
B (0.5)	1	12.5 × 4.5	Hypersil-ODS (5)	E-1	NA	UV (312)	8	—	—	1

Note: E-1 = Methanol + 0.1 *M* sodium acetate.
1 : 1

REFERENCE

1. Dell, D. and Chamberlain, J., *J. Chromatogr.*, 146, 465, 1978.

MORAZONE

Gas Chromatography

Specimen (mℓ)	S	Column m × mm	Packing (mesh)	Oven temp (°C)	Gas (mℓ/min)	Det.	RT (min)	Internal standard (RT)	Deriv.	Other compounds (RT)	Ref.
U (10)	1	1.2 × 4	10% Apiezon® L[a] Chromosorb W (NA)	190	NA (40)	FID	—[b]	*n*-Hexadecane (NA)	—	Phenmetrazine (NA)	1

Thin-Layer Chromatography

Specimen (mℓ)	S	Plate (manufacturer)	Layer (mm)	Solvent	Post-separation treatment	Det. (nm)	R_f	Internal standard (R_f)	Other compounds (R_f)	Ref.
U (10)	3	NA (Merck)	Cellulose (NA)	S-1	Sp: Bromocresol green pH 5.5	Visual	0.7	—	Phenmetrazine (0.5)	1

Note: S-1 = *n*-Butanol + Formic acid + water.
20 : 1 : 1

[a] Alkalinized.
[b] Morazone does not give a peak. Only the peak due to phenmetrazine, the metabolite of morazone present in urine, produces a peak.

REFERENCE

1. Cartoni, G. P., Cavalli, A., Giarusso, A., and Rosati, F., *J. Chromatogr.*, 84, 419, 1973.

MORPHINE

Gas Chromatography

Specimen (mℓ)	S	Column m × mm	Packing (mesh)	Oven temp (°C)	Gas (mℓ/min)	Det.	RT (min)	Internal standard (RT)	Deriv.	Other compounds (RT)	Ref.
U (10)	2	2 × 2	3% OV-17 Gas Chrom® Q (100/120)	230	Methane (19)	MS-CI	3.3	[$^2H_3$]-Morphine (3.3)	Trimethyl-silyl	—	1
U (15)	3	1.8 × 2	3% Poly A-103 Gas Chrom® Q (100/120)	230	NA	FID	4.9	Nalorphine (7.3)	Acetyl	Codeine (3.4) Oxycodone (3.9)	2[a]
D	3	2 × 2	3% Dexil-300 Supelcoport (100/120)	250	Nitrogen (30)	FID	9	Tetracosane (5)	Trimethyl-silyl	Codeine (8) Thebaine (16)	3
U (2)	2	1.2 × 4	3% OV-17 Gas Chrom® Q (100/120)	230	Argon 95 + methane 5 (30)	ECD	1.8	Codeine (2.8)	Heptafluo-robutyryl	—	4
B (1)	1	0.45 × 4	2% OV-17 Diatomite C (100/120)	245	Helium (50)	MS-EI	1.9	[$^2H_3$]-Morphine (1.9)	i. Penta-fluoro-benxyl[b] ii. Trifluo-roacetyl	Nalorphine (2.4)	5
U (10)	3	1.8 × 2	5% SE-52 Chromosorb® G[c] (100/120)	280	Nitrogen (30)	FID	1.7	—	—	—	6
B (0.5)	1	3 × NA	NA	215	Nitrogen (30)	ECD	7.5	Nalorphine (10.8)	Pentafluo-ropro-pionyl	—	12[k]

High-Pressure Liquid Chromatography

Specimen (mℓ)	S	Column cm × mm	Packing (μm)	Elution	Flow rate (mℓ/min)	Det. (nm)	RT (min)	Internal standard (RT)	Other compounds (RT)	Ref.
U (5)	2	25 × 4.6	Partisil® (7)	E-1	2	Fl[d] (Ex: 320, Fl: 436)	4[e]	Dihydromorphine (6.1)[f]	Nalorphine (2.6)[e]	7
D	3	30 × 4	μ-Bondapak® C_{18} (10)	E-2	1.2	UV (254)	5.6	—	Codeine (7.5)[g] Heroin (19.4)	8
D	3	30 × 4	μ-Bondapak® C_{18} (10)	E-3	2	UV (280)	5	—	Dilaudid® (5.8) Nalorphine (7.1) Codeine (8.3) Heroin (15.4)[h] Papaverine (16.7)[h]	9
D	3	30 × 4	μ-Bondapak® C_{18} (10)	E-4	1	UV (254)	4	—	Methylparaben (5) Propylparaben (7.5)	13

Thin-Layer Chromatography

Specimen (mℓ)	S	Plate (manufacturer)	Layer (mm)	Solvent	Post-separation treatment	Det. (nm)	R_f	Internal standard (R_f)	Other compounds (R_f)	Ref.
U (NA)	3	20 × 20 cm (Analtech)	Silica gel G F_{254} (0.25)	S-1	—	Fl[i] (Reflectance) (Ex: 360, Fl: 470)	0.50	—	Amphetamine[j] (0.65)	10
D	3	20 × 20 cm (Analtech)	Silica gel G (0.25)	S-2	E: I_2 vapors	Visual	0.18	—	14-Hydroxymorphine (0.42) 14-Hydroxymorphinone (0.63) 14-Hydroxydihy-	11

MORPHINE (continued)

Thin-Layer Chromatography

Specimen (mℓ)	S	Plate (manufacturer)	Layer (mm)	Solvent	Post-separation treatment	Det. (nm)	R_f	Internal standard (R_f)	Other compounds (R_f)	Ref.
									dromoephinone (0.63) Dihydromorphinone (0.16)	

Note: E-1 = Methanol + 2 *N* NH_4OH + 1 *N* NH_4NO_3.
30 : 20 : 10

E-2 = Methanol + 0.005 *M* *n*-heptanesulfonic acid in water.
40 : 60

E-3 = Methanol + PIC B-7, pH 3.0 (PIC B-7 = heptanesulfonic acid).
35 : 65

E-4 = Methanol + 0.1% monobasic sodium phosphate hydrate containing 5% methanol, pH 4.
60 : 40

S-1 = Chloroform + acetone + methanol + tert. butylamine.
30 : 40 : 10 : 20

S-2 = Ethyl acetate + methanol + conc. ammonia.
86 : 10 : 4

[a] Results are confirmed by analysis using an alternative column (3% OV-25) and by thin-layer chromatography.
[b] Extractive alkylation.
[c] The support was specially deactivated by reaction with benzoyl chloride and pyridine.
[d] Urine extracts are treated with $K_3Fe(CN)_6$ prior to injection into gas chromatograph.
[e] Retention time of mixed dimer with dihydromorphine.
[f] Retention time of dihydromorphine dimer.
[g] Retention data for a number of synthetic morphinone derivatives are given.

[h] Methanol content of the mobile phase was increased from 35% to 55% after 10 min.

[i] The plates were exposed to HCl vapors, sprayed with 2% ferric chloride + 2% potassium ferricyanide (1:1), and then spots were scanned in the transmission mode at 675 nm.

[j] Quantitated after spraying with fluorescamine reagent.

[k] A radioisotopic derivatization procedure is also described. Derivatization is carried out with ^{3}H-dansyl chloride; the dansyl derivative of morphine is separated by thin-layer chromatography prior to determination of radioactivity.

REFERENCES

1. Clarke, P. A. and Foltz, R. L., *Clin. Chem.*, 20, 465, 1974.
2. Jain, N. C., Sneath, T. C., Budd, R. D., and Leung, W. J., *Clin. Chem.*, 21, 1486, 1975.
3. Rasmussen, K. E., *J. Chromatogr.*, 120, 491, 1976.
4. Nicolau, G., van Lear, G., Kaul, B., and Davidow, B., *Clin. Chem.*, 23, 1640, 1977.
5. Cole, W. J., Parkhouse, J., and Yousef, Y. Y., *J. Chromatogr.*, 136, 409, 1977.
6. Street, H. V., Vycudilik, W., and Machata, G., *J. Chromatogr.*, 168, 117, 1979.
7. Jane, I. and Taylor, J. F., *J. Chromatogr.*, 109, 37, 1975.
8. Olieman, C., Maat, L., Waliszewski, K., and Beyerman, H. C., *J. Chromatogr.*, 133, 382, 1977.
9. Soni, S. K. and Dugar, S. M., *J. Forensic Sci.*, 24, 437, 1979.
10. Sherma, J., Dobbins, M. F., and Touchstone, J. C., *J. Chromatogr. Sci.*, 12, 300, 1974.
11. Liras, P., *J. Chromator.*, 106, 238, 1975.
12. Garret, E. R. and Gürkan, T., *J. Pharm. Sci.*, 67, 1512, 1978.
13. Austin, K. L. and Mather, L. E., *J. Pharm. Sci.*, 67, 1510, 1978.

MUZOLIMINE

Thin-Layer Chromatography

Specimen (mℓ)	S	Plate (manufacturer)	Layer (mm)	Solvent	Post-separation treatment	Det. (nm)	R_f	Internal standard (R_f)	Other compounds (R_f)	Ref.
B (1)	1	20 × 20 cm (Merck)	Silica gel 60 (0.25)	S-1	Sp: 1% 4-Dimethylaminocinnamaldehyde in 6 *M* HCl/ethanol (1:1)[a]	Reflectance (610)	0.21	—	—	1

Note: S-1 = Chloroform + ethanol.
9 : 1

a. Ten mℓ of this stock spray reagent is diluted with 80 mℓ of ethanol prior to use.

REFERENCE

1. Ritter, W., *J. Chromatogr.*, 142, 431, 1977.

NADOLOL

Gas Chromatography

Specimen (mℓ)	S	Column m × mm	Packing (mesh)	Oven temp (°C)	Gas (mℓ/min)	Det.	RT (min)	Internal standard (RT)	Deriv.	Other compounds (RT)	Ref.
B (4)	1	1.8 × 3	3% OV-1 Supelcoport (80/100)	Temp. Progr.	Helium (6.5)	MS-EI	6.7	*N*-Methylnadolol (7.4)	Trimethylsilyl	—	1

REFERENCE

1. Funke, P. T., Malley, M. F., Ivashkiv, E., and Cohen, A. I., *J. Pharm. Sci.*, 67, 653, 1978.

NAFIVERINE

Thin-Layer Chromatography

Specimen (mℓ)	S	Plate (manufacturer)	Layer (mm)	Solvent	Post-separation treatment	Det. (nm)	R_f	Internal standard (R_f)	Other compounds (R_f)	Ref.
U (NA)	3	NA	Silica gel G (0.5)	S-1[a] (NA)	Sp: Sat. $K_2Cr_2O_7$ in conc. H_2SO_4	Visual	0.93	—	α-Naphthylpropionic acid (0.71)[b] *N N*-Di (2-hydroxyethyl) piperazine (0.01)[b]	1

Note: S-1 = Acetone + ethyl acetate + chloroform.
3 : 2 : 1

[a] Other solvent systems are also described.
[b] Metabolite of nafiverine is quantitated by scraping and eluting the spot.

REFERENCE

1. Naito, S.-I. and Yamanoto, T., *J. Pharm. Sci.*, 64, 253, 1974.

NALBUPHINE

Gas Chromatography

Specimen (mℓ)	S	Column m × mm	Packing (mesh)	Oven temp (°C)	Gas (mℓ/min)	Det.	RT (min)	Internal standard (RT)	Deriv.	Other compounds (RT)	Ref.
B (3)	1	1.2 × 4	3.8% UCC-W-982 Chromosorb® W (NA)	235	Helium (75)	ECD	3.7	Naloxone (2.2)	Heptafluorobutyryl	—	1

Note: E-1 = Methanol + 0.1 *M* citrate buffer, pH 3.
95 : 15

E-2 = Methanol + water.
63 : 37

REFERENCE

1. **Weinstein, S. H., Alteras, M., and Gaylord, J.,** *J. Pharm. Sci.*, 67, 547, 1978.

NALIDIXIC ACID

Gas Chromatgraphy

Specimen (mℓ)	S	Column m × mm	Packing (mesh)	Oven temp (°C)	Gas (mℓ/min)	Det.	RT (min)	Internal standard (RT)	Deriv.	Other compounds (RT)	Ref.
B (1)	2	1.5 × NA	10% OV-17 Chromosorb® W (80/100)	290	Nitrogen (20)	FID	8	Phenprocoumon[a] (7)	Butyl[b]	—	1
D	3	1.8 × 2	1% OV-1 Chromosorb® W (80/100)	Temp. Progr.	Nitrogen (41)	FID	4.1	α-Cholestane (7.4)	Methyl[c]	—	2

High-Pressure Liquid Chromatography

Specimen (mℓ)	S	Column cm × mm	Packing (μm)	Elution	Flow rate (mℓ/min)	Det. (nm)	RT (min)	Internal standard (RT)	Other compounds (RT)	Ref.
B (2) U (1)	2	25 × 4.6	Partisil® PXS (10/25)	E-1	1.6	UV (254)	—	1,4-Dihydro-4-oxo-1-propyl-1,8-naphthyridine-3,7-dicarboxylic acid (10.5)	1-Ethyl-1,4-dihydro-4-oxo-1,8-naphthyridine-3,7-dicarboxylic acid (12)	3
B (0.1)	2	30 × 4	μ-Bondapak® C_{18} (10)	E-2	1.5	UV (313)	4.2[d]	Propyl nalidixate (7.5)	—	4

[a] Flufenamic acid is also added to minimize losses due to adsorption. It elutes with the solvent front.
[b] Greely's procedure (*Clin. Chem.*, 20, 192, 1974).
[c] With diazomethane.
[d] Nalidixic acid is converted to its methyl ester prior to chromatography with methyl iodide in the presence of cesium carbonate.

REFERENCES

1. Roseboom, H., Sorel, R. H. A., Lingeman, H., and Bouwman, R., *J. Chromatogr.*, 163, 92, 1979.
2. Wu, H. -L., Nakagawa,T., and Uno, T., *J. Chromatogr.*, 157, 297, 1978.
3. Lee, F. H., Koss, R., O'Neil, S. K., Kullberg, M. P., McGarth, M., and Edelson, J., *J. Chromatogr.*, 152, 145, 1978.
4. Sorel, R. H. A. and Roseboom, H., *J. Chromatogr.*, 162, 461, 11979.

NALTREXONE

Gas Chromatography

Specimen (mℓ)	S	Column m × mm	Packing (mesh)	Oven temp (°C)	Gas (mℓ/min)	Det.	RT (min)	Internal standard (RT)	Deriv.	Other compounds (RT)	Ref.
B (0.5—2)	1	1.8 × 2	3% OV-22 Supelcoport (80/100)	215	Argon 90 + Methane 10 (35)	ECD	5.2	Naloxone (3.3)	Pentafluoropropionyl	β-Naltrexol (4.4)	1
D	1[a]	1.8 × 2	3% OV-17 Gas Chrom® Q (100/120)	205	Argon 95 + Methane 5 (40)	ECD	7.5	—	Heptafluorobutyryl	Naloxone (4.5)	2
B (NA) D	1	2.4 × 2	3% OV-17 Gas Chrom® Q (100/120)	205	Argon 95 + Methane 5 (30)	ECD	8	—	Pentafluoropropionyl	Naloxone (5) Naltrexone-3-methyl-ether (14)	3

[a] The procedure is developed for pharmacokinetic studies using pure compounds.

REFERENCES

1. Verbey, K., Kogan, M. J., de Pace, A., and Mule, S. J., *J. Chromatogr.*, 118, 331, 1976.
2. Sams, R. A. and Malspeis, L., *J. Chromatogr.*, 125, 409, 1976.
3. Bruce, G. L., Bhat, H. B., and Sokdoski, T., *J. Chromatogr.*, 137, 323, 1977.

NAPROXEN

Gas Chromatography

Specimen (mℓ)	S	Column m × mm	Packing (mesh)	Oven temp (°C)	Gas (mℓ/min)	Det.	RT (min)	Internal standard (RT)	Deriv.	Other compounds (RT)	Ref.
B (1)	2	1.2 × 3	3.8% SE-30 Diatoport S (80/100)	170	Helium (70)	FID	NA	6-Methoxy-2-naphthyl acetic acid (NA)	Methyl[a]	—	1

High-Pressure Liquid Chromatography

Specimen (mℓ)	S	Column cm × mm	Packing (μm)	Elution	Flow rate (mℓ/min)	Det. (nm)	RT (min)	Internal standard (RT)	Other compounds (RT)	Ref.
B (0.5)	2	30 × 3.9	μ-Bondapak® CN (10)	E-1	2	UV (313)	6	*p*-Chlorowarfarin (12)	Desmethylnaproxen (3.3)	2
B (0.1)	2	30 × 3.9	μ-Bondapak® C_{18} (10)	E-2	0.8	UV (225)	5.7	—	Oxyphenbutazone (5.8) Indomethacin (8) Ibuprofen (8.6) Phenylbutazone (9.6)	3

Note: E-1 = Acetonitrile + 0.5% acetic acid.
30 : 70

E-2 = Acetonitrile + 45 m*M* KH_2PO_4, pH 3.
60 : 40

[a] With diazomethane.

NAPROXEN (continued)

REFERENCES

1. Runkel, R., Chaplin, M., Boost, G., Segre, E., and Forchielli, E., *J. Pharm. Sci.*, 61, 703, 1972.
2. Slattery, J. T. and Levy, G., *Clin. Biochem.*, 12, 100, 1979.
3. Dusci, L. J. and Hackett, L. P., *J. Chromatogr.*, 172, 516, 1979.

NEFOPAM

Gas Chromatography

Specimen (mℓ)	S	Column m × mm	Packing (mesh)	Oven temp (°C)	Gas (mℓ/min)	Det.	RT (min)	Internal standard (RT)	Deriv.	Other compounds (RT)	Ref.
B (5)	2	0.9 × 2	3% OV-225 Gas Chrom® Q (100/120)	170	Nitrogen (25)	FID	9.5	*N*-Isopropyl analog (15)	—	—	1
B (2)	1	1.8 × 2	3% HI EFF 8BP Gas Chrom® Q (100/120)	221	Helium (40)	FID	2.5	Orphenadrine (1)	—	—	2

REFERENCES

1. Ehrsson, H. and Eksborg, S., *J. Chromatogr.*, 136, 154, 1977.
2. Schuppan, D., Hansen, C. S., and Ober, R. E., *J. Pharm. Sci.*, 67, 1720, 1978.

NEOMYCIN

High-Pressure Liquid Chromatography

Specimen (mℓ)	S	Column cm × mm	Packing (μm)	Elution	Flow rate (mℓ/min)	Det. (nm)	RT (min)	Internal standard (RT)	Other compounds (RT)	Ref.
D	3	25 × 4.6	LiChrosorb® SI-100 (5)	E-1	1.0	Visible[a] (350)	B[b] = 14 C = 9.9	—	Neamine (2.1) Deoxystreptamine (12.7) Neobiosamine (23.6) *N*-Acetylneomycin (26.6)	1[c]

Note: E-1 = Chloroform + tetrahydrofuran + water.
600 : 392 : 8

[a] 2,4-Dinitrobenzene derivatives are prepared prior to chromatography.
[b] Neomycin is a mixture of two stereoisomers, neomycin B and C.
[c] The authors claim that their earlier gas-chromatography procedure (*J. Pharm. Sci.*, 60, 1068, 1971) is a complicated method to perform.

REFERENCE

1. Tsuji, K., Goetz, J. F., van Meter, W., and Gusciora, K. A., *J. Chromatogr.*, 175, 141, 1979.

NEOSTIGMINE

Gas Chromatography

Specimen (mℓ)	S	Column m × mm	Packing (mesh)	Oven temp (°C)	Gas (mℓ/min)	Det.	RT (min)	Internal standard (RT)	Deriv.	Other compounds (RT)	Ref.
B (3)	1	2 × NA	3% OV-17 Chromosorb® W (100/120)	200	Nitrogen (30)	NPD	8.7	Pyridostigmine (2.6)	a	—	1
U (3)	2	2 × NA	10% OV-17 Chromosorb® W (100/120)	200	Nitrogen (30)	NPD	—	3-Hydroxy-*N*-methylpyridinium bromide (2.2)	a	3-Hydroxyphenyltrimethylammonium bromide (6.3)	2

[a] Neostigmine and pyridostigmine were extracted as glycine-iodide complexes with dichloromethane.

REFERENCES

1. Chan, K., Williams, N. E., Baty, J. D., and Calvey, T. N., *J. Chromatogr.*, 120, 349, 1976.
2. Chan, K. and Deghan, A., *J. Pharmacol. Methods*, 1, 311, 1978.

NETILMICIN

High-Pressure Liquid Chromatography

Specimen (mℓ)	S	Column cm × mm	Packing (μm)	Elution	Flow rate (mℓ/min)	Det. (nm)	RT (min)	Internal standard (RT)	Other compounds (RT)	Ref.
B (0.1)	2	30 × 4	μ-Bondapak® C_{18} (10)	E-1	1	Fl[a] (Ex: 220, Fl: 470[b])	5.8	—	—	1
B (0.2)	2	30 × 4	μ-Bondapak® C_{18} (10)	E-2	2	Fl[c] (d)	9	—	Tobramycin (8)	2

Note: E-1 = Acetonitrile + water.
95 : 5

E-2 = Methanol + triethylamine + 0.5 *M* Tris buffer, pH 7.9.
74 : 1 : 25

[a] Dansyl derivative is prepared prior to chromatography.
[b] Filter KV-470.
[c] Derivatization is carried out with *o*-phthalaldehyde prior to chromatography.
[d] Ex. filters 7-54 and 7-60; Fl. filters 4-76 and 3-72 (Varian).

REFERENCES

1. **Peng, G. W., Jackson, G. G., and Chiou, W. L.,** *Antimicrob Agents Chemother.*, 12, 707, 1977.
2. **Bäck, S.-E., Nilsson-Ehle, I., and Nilsson-Ehle, P.,** *Clin. Chem.*, 25, 1222, 1979.

NICOTINE

Gas Chromatography

Specimen (mℓ)	S	Column m × mm	Packing (mesh)	Oven temp (°C)	Gas (mℓ/min)	Det.	RT (min)	Internal standard (RT)	Deriv.	Other compounds (RT)	Ref.
B (1)	1	1.8 × 2	3% SP-2250DB Supelcoport (100 / 120)	155	Helium (30)	NPD	1.9	Modaline (2.9)	—	a	1
B (3)	1	20 × 0.3[b]	SP-1000	160	Helium (0.5)	MS-EI	2.4	Quinoline (3)	—	—	2
B,U (3)	1	1.8 × NA	10% Apiezon L + 10% KOH Chromosorb® W (80/100)	220	Helium (60)	NPD	3	Quinoline (2.5)	—	—	3
Breast fluid (0.05)	1	2 × 2	2% Carbowax® + 2% KOH Chromosorb® P (NA)	Temp. Progr.	Helium (25)	MS-EI	4.5	[$^2H_2$]-Nicotine (4.5)	—	—	4[c]
B (1)	2	2.1 × 2	3% OV-17 Gas Chrom® Z (80/100)	225	Nitrogen (30)	ECD	7.9[d]	*N*-*n*-Propylnor-nicotine (9.8)[d]	e	—	5

High-Pressure Liquid Chromatography

Specimen (mℓ)	S	Column cm × mm	Packing (μm)	Elution	Flow rate (mℓ/min)	Det. (nm)	RT (min)	Internal standard (RT)	Other compounds (RT)	Ref.
U (3)	2	25 × 2	MicroPak SI-10 (10)	E-1	0.8	UV (260)	1.4	Desmethylimi-pramine (4.2)	Cotinine (2.1)	6
U (20)	2	25 × 4.6	Zorbax Sil® (NA)	E-2	1.0	UV (254)	4.5	—	Nornicotine (4.2) Cotinine (10)	7

NICOTINE (continued)

Note: E-1 = Ethyl acetate + 2-propanol + ammonia.
80 : 3 : 0.4

E-2 = Dioxane + 2-propanol + ammonia.

[a] Conditions for the analysis of cotinine are also described.
[b] Wall-coated, open-tubular column.
Problems of contamination of laboratory air with the assay of nicotine are discussed.
[d] Extra peaks are obtained due to the on-column decomposition of the derivatized nicotine and the internal standard.
[e] 8-Chlorotrichloroethyl-carbamate of nicotine and of the internal standard are formed on treatment with trichloroethylchloroformate.

REFERENCES

1. **Hengen, N. and Hengen, M.,** *Clin. Chem.,* 24, 50, 1978.
2. **Dow, J. and Hall, K.,** *J. Chromatogr.,* 153, 521, 1978.
3. **Feyerabend, C. and Russel, M. A. H.,** *J. Pharm. Pharmacol.,* 31, 73, 1979.
4. **Gruenke, L. D., Beelen, T. C., Craig, J. C., and Petrakis, N. L.,** *Anal. Biochem.,* 94, 411, 1979.
5. **Hartvig, P., Ahnfelt, N. -O., Hammerlund, M., and Vessman, J.,** *J. Chromatogr.,* 173, 127, 1979.
6. **Watson, I. D.,** *J. Chromatogr.,* 143, 203, 1977.
7. **Maskarinec, M. P., Harvey, R. W., and Caton, J. E.,** *J. Anal. Toxicol.,* 2, 124, 1978.

NICOTINIC ACID

Gas Chromatography

Specimen (mℓ)	S	Column m × mm	Packing (mesh)	Oven temp (°C)	Gas (mℓ/min)	Det.	RT (min)	Internal standard (RT)	Deriv.	Other compounds (RT)	Ref.
D	3	1.5 × 1.8	20% Carbowax® 20 *M* Terphthalic acid Gas Chrom® P (100/120)	155	Nitrogen (30)	FID	—	6-Methylnicotinamide (6.5)	a	Nicotinamide (5.5)	1

High-Pressure Liquid Chromatography

Specimen (mℓ)	S	Column cm × mm	Packing (μm)	Elution	Flow rate (mℓ/min)	Det. (nm)	RT (min)	Internal Standard (RT)	Other compounds (RT)	Ref.
D	3	100 × 2.1	Zipax® SCX (NA)	E-1	b	UV (254)	2.5	—	Riboflavin (3.5) Pyridoxine (6.5)	2[c]
D	3	30 × 4	μ-Bondapak® C_{18} (10)	E-2	3	UV (254)	3.5	—	Nicotinamide (3)	3
B (0.5)	2	30 × 4	μ-Bondapak® C_{18} (10)	E-3	2	UV (254)	7.8	Isonicotinic acid (6.8)	Nicotinuric acid (8.4)	4
U (0.003)	2	25 × 4.6	Partisil®-10 ODS (10)	E-4	1.5	UV (254	—	—	*N*-Methylnicotinamide (16)	5

Note: E-1 = 0.5 *M* Phosphate buffer, pH 4.4.

E-2 = Methanol + 0.002 *M* dioctylsodium sulfosuccinate, pH 2.5.
50 : 50

E-3 = Water + methanol + tetrabutylammonium phosphate.
90 : 100 : (50 mmoles)

NICOTINIC ACID (continued)

E-4 = Methanol + 0.005 *M* sodium dodecylsulfate + sulfuric acid.
47.4 : 52.5 : 0.01

[a] The amides are converted to corresponding nitriles by treatment with trifluoroacetic anhydride.
[b] Column pressure = 1500 psi.
[c] Conditions for separation of other water soluble vitamins are also described.

REFERENCES

1. **Vessman, J. and Strömberg, S.,** *J. Pharm. Sci.*, 64, 311, 1975.
2. **Williams, R. C., Baker, D. R., and Schmidt, J. A.,** *J. Chromatogr. Sci.*, 11, 618, 1973.
3. **Sood, S. P., Wittmer, D. P., Ismael, S. A., and Haney, W. G.,** *J. Pharm. Sci.*, 66, 40, 1977.
4. **Hengen, N., Seiberth, V., and Hengen, M.,** *Clin. Chem.*, 24, 1740, 1978.
5. **Shaikh, B. and Pontzer, N. J.,** *J. Chromatogr.*, 162, 596, 1979.

NIFEDIPINE

Gas Chromatography

Specimen (mℓ)	S	Column m × mm	Packing (mesh)	Oven temp (°C)	Gas (mℓ/min)	Det.	RT (min)	Internal standard (RT)	Deriv.	Other compounds (RT)	Ref.
B (1)	1	1 × 3	3% OV-1 Chromosorb® W (80/100)	200	Helium (30)	MS-EI	2	[^{2}H]-Nifedipine (2)	Oxidation[a]	—	1
B (0.5)	1	1.8 × 2	2% OV-17 Gas Chrom® Q (80/100)	240	Nitrogen (25)	ECD	4	Diazepam (2.4)	—	Metabolite[b] (1.3)	2

[a] 1.4-Dihydropyridine ring of nifedipine is oxidized to the corresponding pyridine on treatment with aqueous sodium nitrite solution.
[b] The metabolite is identical with the oxidation product of nifedipine.

REFERENCES

1. **Higuchi, S. and Shiobara, Y.,** *Biomed. Mass Spectrom.*, 5, 220, 1978.
2. **Jakobsen, P., Pedersen, O. L., and Mikkelsen, E.,** *J. Chromatogr.*, 162, 81, 1979.

NIKETHAMIDE

Gas Chromatography

Specimen (mℓ)	S	Column m × mm	Packing (mesh)	Oven temp (°C)	Gas (mℓ/min)	Det.	RT (min)	Internal standard (RT)	Deriv.	Other compounds (RT)	Ref.
B (1)	1	NA	3% OV-1 Gas Chrom® Q (NA)	155	Nitrogen (40)	FID	2.5	Phenacetin (5)	—	*N*-Ethylnicotinamide (1.9)	1

Thin-Layer Chromatography

Specimen (mℓ)	S	Plate (manufacturer)	Layer (mm)	Solvent	Post-separation treatment	Det. (nm)	R_f	Internal standard (R_f)	Other compounds (R_f)	Ref.
U	3	20 × 20 cm (Laboratory)	Silica gel G F_{254} (0.25)	S-1	Sp: KI (1g) + I_2 (1g) + ethanol (25 mℓ) + conc HCl (6 mℓ) + water to make 50 mℓ	Visual	0.70	—	*N*-Ethylnicotinamide (0.57)	1

Note: S-1 = Ethyl acetate + acetone + chloroform + ammonia.
100 : 10 : 5 : 5

REFERENCE

1. Lewis, J. H., *J. Chromatogr.*, 172, 295, 1979.

NIMORAZOLE

High-Pressure Liquid Chromatography

Specimen (mℓ)	S	Column cm × mm	Packing (μm)	Elution	Flow rate (mℓ/min)	Det. (nm)	RT (min)	Internal standard (RT)	Other compounds (RT)	Ref.
B (NA)	1	30 × 3.9	μ-Bondapak® C_{18} (10)	E-1	2	UV (313)	4.9	RO 07-0269[a] (3)	—	1

Note: E-1 = 5 m*M* Heptanesulfonic acid in 30% methanol, pH 3.2.

[a] 1-(2-Nitroimidazol-1-y1)-3-chloropropan-2-01.

REFERENCE

1. **Workman, P.**, *J. Chromatogr.*, 163, 396, 1979.

NIRIDAZOLE

High-Pressure Liquid Chromatography

Specimen (mℓ)	S	Column cm × mm	Packing (μm)	Elution	Flow rate (mℓ/min)	Det. (nm)	RT (min)	Internal standard (RT)	Other compounds (RT)	Ref.
B, U (1)	2	25 × 4.6	Spherisorb ODS (5)	E-1	0.5	UV (365)	7.5	—	—	1

Note: E-1 = Methanol + water.
4 : 1

REFERENCE

1. Miller, J. J, Jones, B. M., Massey, P. R., and Salaman, J. R., *J. Chromatogr.*, 147, 507, 1978.

NITRAZEPAM

Gas Chromatography

Specimen (mℓ)	S	Column m × mm	Packing (mesh)	Oven temp (°C)	Gas (mℓ/min)	Det.	RT (min)	Internal standard (RT)	Deriv.	Other compounds (RT)	Ref.
B (1)	2	1.8 × 4	3% OV-17 Gas Chrom® Q (60/80)	235	Argon (50)	ECD	6.1[a]	Clonazepam (8.5)[a]	Hydrolysis	—	1
B (0.5)	2	0.9 × 2	5% OV-17 Gas Chrom® Q (80/100)	250	Nitrogen (30)	ECD	7.4	Griseofulvin (12.1)	Methyl[b]	Metabolite 1[c] (11.7) Metabolite 2[d] (37)	2
B (1)	1	1.5 × 6	3% OV-17 Diatomite CQ (80/100)	245	Argon 90 + Methane 10 (100)	ECD	2.9	*N*-Desmethyldiazepam (4)[e]	Hydrolysis	—	3
B (1)	1	10 × 0.4[f]	3% OV-17 Cab-O-Sil (Grade M5)	230	Argon 95 + Methane 5 (10)	ECD	4	Clonazepam (5)	—	—	4
U (1)	2	0.6 × 2	3% OV-101 Chrmosorb® Q (100/120)	275	Nitrogen (40)	NPD	3.5	Methylnitrazepam (2.5)	—	Metabolite 1[c] (4) Metabolite 2[d] (9)	5[g]

High-Pressure Liquid Chromatography

Specimen (mℓ)	S	Column cm × mm	Packing (μm)	Elution	Flow rate (mℓ/min)	Det. (nm)	RT (min)	Internal standard (RT)	Other compounds (RT)	Ref.
U (10)	3	50 × 2	Zipax® SAX (30)	E-1	1	UV (260)	2.5	—	Metabolite 1[c] (5) Metabolite 2[d] (10.5)	6

Note: E-1 = Ethyl acetate + hexane.
3 : 7

NITRAZEPAM (continued)

[a] Under the same chromatographic conditions, the retention times of intact nitrazepam and clonazepam are 25.5 and 32 min, respectively.
[b] Extractive alkylation with methyl iodide in the presence of tetrabutylammonium hydrogen sulfate.
[c] 7-Amino analog.
[d] 7-Acetamido analog.
[e] Retention time of intact desmethyldiazepam. It is added just prior to injection and escapes hydrolysis step.
[f] Support-coated, open-tubular column.
[g] Urinary metabolites have also been analyzed using an electron capture detector.

REFERENCES

1. **Beharrell, G. P., Hailey, D. M., and McLaurin, M. K.,** *J. Chromatogr.*, 70, 45, 1972.
2. **Ehrsson, H. and Tilly, A.,** *Anal. Lett.*, 6, 197, 1973.
3. **Jensen, K. M.,** *J. Chromatogr.*, 111, 389, 1975.
4. **de Boer, A. G., Röst-Kaiser, J., Bracht, H., and Breimer, D. D.,** *J. Chromatogr.*, 145, 105, 1978.
5. **Kangas, L.,** *J. Chromatogr.*, 172, 273, 1979.
6. **Moore, B., Nickless, G., Hallet, C., and Howard, A. G.,** *J. Chromatogr.*, 137, 215, 1977.

NITROFURANTOIN

High-Pressure Liquid Chromatography

Specimen (mℓ)	S	Column cm × mm	Packing (μm)	Elution	Flow rate (mℓ/min)	Det. (nm)	RT (min)	Internal standard (RT)	Other compounds (RT)	Ref.
B, U (0.2)	2	30 × 3.9	μ-Bondapak® C_{18} (10)	E-1	2	Absorbance (365)	6	Furazolidone (8)	—	1
B (0.01)	1	15 × 4.6	LiChrosorb® RP-8 (5)	E-2	1.6	Absorbance (370)	8	—	Metronidazole (5.5) Hydroxymethyl-nitrofurantoin (8) Nitrofural (9.9) Nitrofurazolidine (10.4)	2

Note: E-1 = Methanol + .01 *M* sodium acetate, pH 5.0.
20 : 80

E-2 = Ethanol + water.
5 : 5

REFERENCES

1. Aufrère, M. B., Hoener, B. -A., and Vore, M. E., *Clin. Chem.*, 23, 2207, 1977.
2. Vree, T. B., Hekster, Y. A., Baars, A. M., Damsma, J. E., van der Kleijn, E., and Bron J., *J. Chromatogr.*, 162, 110, 1979.

NITROGLYCERIN

Gas Chromatography

Specimen (mℓ)	S	Column m × mm	Packing (mesh)	Oven temp (°C)	Gas (mℓ/min)	Det.	RT (min)	Internal standard (RT)	Deriv.	Other compounds (RT)	Ref.
B (0.2)	2	NA	3% SP-2401 Supelcoport (100/120)	140	Nitrogen (60)	ECD	6	Isosorbide dinitrate (13)	—	—	1
B (5)	2	1.8 × 2	30% SE-30 Chromosorb® W (80/100)	130	Argon 90 + Methane 10 (90)	ECD	11	*m*-Dinitrobenzene (19)	—	—	2

High-Pressure Liquid Chromatography

Specimen (mℓ)	S	Column cm × mm	Packing (μm)	Elution	Flow rate (mℓ/min)	Det. (nm)	RT (min)	Internal standard (RT)	Other compounds (RT)	Ref.
D	3	30 × 3.9	μ-Bondapak®-phenyl (10)	E-1	2	UV(218)	9	Isosorbide dinitrate (6)	—	3

Note: E-1 = Acetonitrile + tetrahydrofuran + water.
260 : 100 : 640

REFERENCES

1. **Yap, P. S. K., McNiff, E. F., and Fung, H. -L.,** *J. Pharm. Sci.*, 67, 582, 1978.
2. **Wei, J. Y. and Reid, P. R.,** *Circulation*, 59, 588, 1979.
3. **Baaske, D. M., Carter, J. E., and Amann, A. H.,** *J. Pharm. Sci.*, 68, 481, 1979.

NOMIFENSINE

Gas Chromatography

Specimen (mℓ)	S	Column m × mm	Packing (mesh)	Oven temp (°C)	Gas (mℓ/min)	Det.	RT (min)	Internal standard (RT)	Deriv.	Other compounds (RT)	Ref.
B (2)	1	2 × 4	3% OV-17 Gas Chrom® Q (100/120)	245	Nitrogen (40)	ECD	NA	2-Amino-5-chlorobenzophenone (NA)	Heptafluorobutyryl	—	1
B (2)	1	50 × 0.25[a]	OV-101	Temp. Progr.	Helium (2)	NPD	15	2-Propyl analog (17.5) + 2-Butyl analog (20)	Heptafluorobutyryl	—	2
B (1)	1	2 × NA	3% OV-25 Chromosorb® W (NA)	250	Helium (30)	NPD	3	HOE-49673[b] (4.5)	—	—	3

Thin-Layer Chromatography

Specimen (mℓ)	S	Plate (manufacturer)	Layer (mm)	Solvent	Post-separation treatment	Det. (nm)	R_f	Internal standard (R_f)	Other compounds (R_f)	Ref.
B (1.5)	2	20 × 20 cm Alumium sheets (Merck)	Silica gel 60 (0.2)	S-I	D: Perchloric acid (50 mℓ) in water (60 mℓ) and ethanol (60 mℓ)	Fl (Ex: UVB filter, Fl: 538 nm)	0.75	—	—	4

Note: S-1 = Dioxane + benzene + acetone + ammonia.
20 : 10 : 6 : 4

[a] Wall coated open tubular column.
[b] Nomifensine analog.

NOMIFENSINE (continued)

REFERENCES

1. Vereczkey, L., Bianchetti, G., Rovei, V., and Frigerio, A., *J. Chromatogr.*, 116, 451, 1976.
2. Bailey, E., Fenoughty, M., and Richardson, L., *J. Chromatogr.*, 131, 347, 1977.
3. Chamberlain, J. and Hill, H. M., *Br. J. Clin. Pharmacol.*, 4, 117 S, 1977.
4. Klitgaard, N. A., *Arch. Pharm. Chemi. Sci. Ed.*, 6, 29, 1978.

NORTRIPTYLINE

Gas Chromatography

Specimen (mℓ)	S	Column m × mm	Packing (mesh)	Oven temp (°C)	Gas (mℓ/min)	Det.	RT (min)	Internal standard (RT)	Deriv.	Other compounds (RT)	Ref.
B (0.5) CSF	1	1.2 × 1.5	0.75% OV-17 Chromosorb® W (100/120)	190—220	Helium (20)	MS-EI	5	—	Trifluoroacetyl	10,11-Dehydronortriptyline (5.8)	1
B, U (4)	2	3.2 × 1.8	0.75% OV-17 Chromosorb® G (80/100)	245	Nitrogen (15)	ECD	3.7	Maprotiline (5.9)	Heptafluorobutyryl	10,11-Dehydronortriptyline (4.5)	2
B (2)	2	1.8 × 2	2% SE-30 Gas Chrom® Q (80/00)	215	Helium (50)	a	3.2	—	Acetylation[b]	—	3
B (1)	2	1.8 × 4	3% SP-2250 NA	250	Argon 95 + Methane 5 (60)	ECD	3.2	Maprotiline (4.8)	Heptafluorobutyryl	M —	4
B (1)	1	1.2 × 2	3% OV-17 Gas Chrom® Q (120/140)	240	Isobutane (NA)	MS-CI	1.5	[$^2H_3$]-Nortriptyline (1.5) + [$^2H_6$]-Amitriptyline (1.5)[c]	Trifluoroacetyl	Amitriptyline (1.5)[c] 10,11-Dehydroamitriptyline (1.7)[c] 10,11-Dehydronortriptyline (1.7)	5

High-Pressure Liquid Chromatography

Specimen (mℓ)	S	Column cm × mm	Packing (μm)	Elution	Flow rate (mℓ/min)	Det. (nm)	RT (min)	Internal standard (RT)	Other compounds (RT)	Ref.
D	3	100 × 2.1	Corasil-phenyl (37—50)	E-1	1.0	UV (254)	6	Triflupromazine (3)	—	6

NORTRIPTYLINE (continued)

Thin-Layer Chromatography

Specimen (mℓ)	S	Plate (manufacturer)	Layer (mm)	Solvent	Post-separation treatment	Det. (nm)	R_f	Internal standard (R_f)	Other compounds (R_f)	Ref.
B (1)	2	5 × 20 cm (Analtech)	Silica gel GF (NA)	S-1[d]	—	e	0.44	—	—	7

Note: E-1 = Methanol + acetonitrile + 0.25% ammonium carbonate.
40 : 40 : 20

S-1 = Cyclohexane + 1,4-dioxane.
1 : 1

[a] Gas proportional to counter to measure radioactivity.
[b] With ^{3}H-acetic anhydride.
[c] Analysis carried out at different temperature.
[d] Plasma extracts are treated with (^{14}C) acetic anhydride prior to chromatography.
[e] Detection of radioactivity with a radiochromatoscanner.

REFERENCES

1. Hamma, C.G., Alexanderson, B., Holmstedt, B., and Sjoqvist, F., *Clin. Pharmacol. Ther.*, 12, 496, 1971.
2. Borga, O. and Garle, M., *J. Chromatogr.*, 68, 77, 1972.
3. Loh, A., Zuleski, F R., and DiCarlo, F. J., *J. Pharm. Sci.*, 66, 1056, 1977.
4. Robinson, J. D., Braithwaite, R. A., and Dawling, S., *Clin. Chem.* 24, 2023, 1978.
5. Garland, W. A., Muccino, R. R., Min, B. H., Cupano, J., and Fann, W. E., *Clin. Pharmacol. Ther.*, 25, 844, 1979.
6. Salmon, J. R. and Wood, P. R., *Analyst (London)*, 101, 611, 1976.
7. Zuleski, F. R., Loh, A., and DiCarlo, F. J., *J. Chromatogr.*, 132, 45, 1977.

NOVOBIOCIN

High-Pressure Liquid Chromatography

Specimen (mℓ)	S	Column cm × mm	Packing (μm)	Elution	Flow rate (mℓ/min)	Det. (nm)	RT (min)	Internal standard (RT)	Other compounds (RT)	Ref.
D	3	100 × 2.1	Zipax® HCP (NA)	E-1	0.85	UV (254)	24	Prednisolone (8.5)	Descarbamylnovobiocin (13) Isonovobiocin (19)	1

Note: E-1 = Methanol + 0.02 *M* phosphate buffer, pH 7.
15 : 85

REFERENCE

1. **Tsuji, K. and Robertson, J. H.,** *J. Chromatogr.*, 94, 245, 1974.

NYLIDRIN

Gas Chromatography

Specimen (mℓ)	S	Column m × mm	Packing (mesh)	Oven temp (°C)	Gas (mℓ/min)	Det.	RT (min)	Internal standard (RT)	Deriv.	Other compounds (RT)	Ref.
U (5)	2	1.8 × NA	0.2% OV-1 GLC-110 glass beads (80/100)	Temp. Progr.	Helium (50)	FID	11.7	*n*-Docosane (4.7)	Trimethyl-silyl	—	1

REFERENCE

1. Li, H. and Cervoni, P., *J. Pharm. Sci.*, 65, 1352, 1976.

OLAQUINDOX

High-Pressure Liquid Chromatography

Specimen (mℓ)	S	Column cm × mm	Packing (μm)	Elution	Flow rate (mℓ/min)	Det. (nm)	RT (min)	Internal standard (RT)	Other compounds (RT)	Ref.
Feeds	3	25 × 2.1	Spherisorb ODS- C_{18} (5)	E-1	1.2	UV (254)	3	—	—	1

Note: E-1 = Methanol + water.
5 : 95

REFERENCE

1. Bories, G. F., *J. Chromatogr.*, 172, 505, 1979.

ORG-6001

Thin-Layer Chromatography

Specimen (mℓ)	S	Plate (manufacturer)	Layer (mm)	Solvent	Post-separation treatment	Det. (nm)	R_f	Internal standard (R_f)	Other compounds (R_f)	Ref.
B (1)	1	20 × 20 cm (Merck)	Silica gel 60 (0.5)	S-1	D: Ethanol + water + perchloric acid (135 + 135 + 12), heat at 90° for 35 min	Fl (reflectance) (Ex: 345, Fl: 430)	0.22	—	—	1

Note: S-1 = Chloroform + methanol + diethyl ether + 25% ammonia.
30 : 15 : 5 : 0.25

REFERENCE

1. Søndergaard, I. and Steiness, E., *J. Chromatogr.*, 162, 422, 1979.

ORPHENADRINE

Gas Chromagography

Specimen (mℓ)	S	Column m × mm	Packing (mesh)	Oven temp (°C)	Gas (mℓ/min)	Det.	RT (min)	Internal standard (RT)	Deriv.	Other compounds (RT)	Ref.
B (0.5—2)	1	1.3 × 2.3	3% Carbowax® + 3% KOH Gas Chrom® Q (100/120)	200	Helium (30)	NPD	2.7	Diphenhydramine (3.2)	—	*N*-Demethylorphenadrine (4.9) *N,N*-Didemethylorphenadrine (6.2)	1

REFERENCE

1. Labout, J. J. M., Thijssen, C. T., and Hespe, W., *J. Chromatogr.*, 144, 201, 1977.

OXAZEPAM

Gas Chromatography

Specimen (mℓ)	S	Column m × mm	Packing (mesh)	Oven temp (°C)	Gas (mℓ/min)	Det.	RT (min)	Internal standard (RT)	Deriv.	Other compounds (RT)	Ref.
B (2)	1	1.5 × 1.8	3% OV-225 Chromosorb® G (100/120)	265	Nitrogen (30)	ECD	9.1	Lorazepam (12.4)	Methyl[a]	—	1
B (1)	1	1.8 × 2	3% OV-7 Chromosorb® W (80/100)	250	Argon 95 + Methane 5 (40)	ECD	3.5[b]	N-Desmethyldiazepam (5.8)	—	—	2
B, U (1)	1	1 × 3	3% OV-1 Chromosorb® W (80/100)	210	Helium (30)	MS-EI	1.7	Lorazepam (2.7)	Trimethylsilyl	—	3

[a] Extractive alkylation with methyl iodide in the presence of tetrahexylammonium hydrogen sulfate.
[b] Retention time of 6-chloro-4-phenylquinazoline-2-carboxaldehyde, the thermal decomposition product of oxazepam.

REFERENCES

1. Vessman, J., Johansson, M., Magnusson, P., and Strömberg, S., *Anal. Chem.*, 49, 1545, 1977.
2. Giles, H. G., Fan, T., Naranjo, C. A., and Sellers, E. M., *Can. J. Pharm. Sci.*, 13, 64, 1978.
3. Higuchi, S., Urabe, H., and Shiobara, Y., *J. Chromatogr.*, 164, 55, 1979.

γ-OXO-PHENYLBUTAZONE

Gas Chromatography

Specimen (mℓ)	S	Column m × mm	Packing (mesh)	Oven temp (°C)	Gas (mℓ/min)	Det.	RT (min)	Internal standard (RT)	Deriv.	Other compounds (RT)	Ref.
B (1)	1	1.8 × 2.5	3% OV-11 Chromosorb® W (80/100)	285	Nitrogen (70)	FID	2.6[a]	Acenocoumanol (7.1)	Methyl[b]	—	1

[a] A second peak is also shown at 1.5 min. However, only the peak at 2.6 min is used for quantitation.
[b] With diazomethane.

REFERENCE

1. Midha, K. K., McGilveray, I. J., and Cooper, J. K., *J. Pharm Sci.*, 67, 279, 1978.

OXPRENOLOL

Gas Chromatoraphy

Specimen (mℓ)	S	Column m × mm	Packing (mesh)	Oven temp (°C)	Gas (mℓ/min)	Det.	RT (min)	Internal standard (RT)	Deriv.	Other compounds (RT)	Ref.
B (2)	2	1.5 × 2	3% JXR Chromosorb® G (80/100)	200	Nitrogen (30)	ECD	4	Metoprolol (7)	Trifluoroacetyl	—	1
B (0.2)	1	25 × 0.25[a]	OV-101	205	Nitrogen b	ECD	7.5	Alprenolol (6)	Heptafluorobutyryl	—	2

[a] Wall-coated, open-tubular column.
[b] Linear velocity: 58 cm/sec.

REFERENCES

1. Degen, P. H. and Riess, W., *J. Chromatogr.*, 121, 72, 1976.
2. DeBruney, D., Kinsun, H., Moulin, M. A., and Bigot, M. C., *J. Pharm. Sci.*, 68, 511, 1979.

OXYCODONE

Gas Chromatography

Specimen (mℓ)	S	Column m × mm	Packing (mesh)	Oven temp (°C)	Gas (mℓ/min)	Det.	RT (min)	Internal standard (RT)	Deriv.	Other compounds (RT)	Ref.
U (1)	2	1.8 × 2	2% OV-1 Chromosorb® G (100/120)	220	Argon 95 + Methane 5 (30)	ECD	3.8	Naloxone (4.2)	Heptafluorobutyryl	Oxymorphone (3.3)	1
B	1	0.9 × 2	2% OV-101 Chromosorb® W (100/120)	240	Helium (30)	NPD	3	Conydine (6.5)	—	—	2

Thin-Layer Chromatography

Specimen (mℓ)	S	Plate (manufacturer)	Layer (mm)	Solvent	Post-separation treatment	Det. (nm)	R_f	Internal standard (R_f)	Other compounds (R_f)	Ref.
U (NA)	3	NA	Silica gel G (0.25)	S-1	Sp: Iodoplatinate reagent	Visual	0.57	—	Oxymorphone (0.45)	1

Note: S-1 = i. Ethyl acetate + methanol + ammonia (1st development to 7.5 cm).
82 : 13 : 5
ii. Ethyl acetate + methanol (2nd development to 15 cm).
98 : 2

REFERENCES

1. **Baselt, R. C. and Stewart, C. B.,** *J. Anal. Toxicol.*, 2, 107, 1978.
2. **Benzi, N. L. Jr., and Tam, J. N.,** *J. Pharm. Sci.*, 68, 43, 1979.

OXYPHENBUTAZONE

Gas Chromatography

Specimen (mℓ)	S	Column m × mm	Packing (mesh)	Oven temp (°C)	Gas (mℓ/min)	Det.	RT (min)	Internal standard (RT)	Deriv.	Other compounds (RT)	Ref.
B (1)	1	1.2 × 2	3% OV-17 Gas Chrom® W (80/100)	Temp. Progr.	Helium (30)	NPD	4.9	5-(4-Hydroxy-phenyl)-5-phenylhydantoin (5.4)	Trifluoroacetyl	—	1

REFERENCE

1. Bertrand, M., Dupuis, C., Gagnon, M. -A., and Dugal, R., *J. Chromatogr.*, 171, 377, 1979.

OXYPHENONIUM BROMIDE

Gas Chromatography

Specimen (mℓ)	S	Column m × mm	Packing (mesh)	Oven temp (°C)	Gas (mℓ/min)	Det.	RT (min)	Internal standard (RT)	Deriv.	Other compounds (RT)	Ref.
B (1—2)	1	1.8 × 2	3% OV-17 Chromosorb® W (80/100)	235	Argon 95 + Methane 5 (30)	ECD	4	Benactyzine methiodide (5.5)	Pentafluorobenzyl[a]	—	1

[a] The drug is extracted as ion pair with perchlorate, reextracted with tetrapentylammonium as the counter ion, hydrolyzed to cyclohexylphenylglycolic acid, and then derivatized to prepare pentafluorobenzyl ester.

REFERENCE

1. Greving, J. E., Jonkman, J. H. G., Fiks, F., de Zeeuw, R. A., van Bork, L. E., and Orie, N. G. M., *J. Chromatogr.*, 142, 611, 1977.

OXYTOCIN

High-Pressure Liquid Chromatography

Specimen (mℓ)	S	Column cm × mm	Packing (μm)	Elution	Flow rate (mℓ/min)	Det. (nm)	RT (min)	Internal standard (RT)	Other compounds (RT)	Ref.
D	3	25 × 4	Nuclosil C_{18}[a] (10)	E-1	4.0	UV (210)	4	—	b	1

Note: E-1 = Acetonitrile + 4/15 *M* phosphate buffer, pH 7.
1 : 4

[a] Other packings have also been used.
[b] Conditions for the analysis of other nonapeptides, e.g., demoxytocin and ornipressin, have also been described.

REFERENCE

1. Krummen, K. and Frei, R. W., *J. Chromatogr.*, 132, 429, 1977.

PANCURONIUM BROMIDE

Thin-Layer Chromatography

Specimen (mℓ)	S	Plate (manufacturer)	Layer (mm)	Solvent	Post-separation treatment	Det. (nm)	R_f	Internal standard (R_f)	Other compounds (R_f)	Ref.
D	3	20 × 20 cm (Woelm)	Silica gel (0. 25)	S-1	E: I_2 vapor	Visual	0.2	—	a	1

Note: S-1 = 1-Butanol + pyridine + acetic acid + 20% ammonium chloride (upper phase).
60 : 40 : 12 : 48

[a] R_f values of different hydrolysis products are given.

REFERENCE

1. Kinget, R. and Michoel, A., *J. Chromatogr.*, 120, 234, 1976.

PAPAVERINE

Gas Chromatography

Specimen (mℓ)	S	Column m × mm	Packing (mesh)	Oven temp (°C)	Gas (mℓ/min)	Det.	RT (min)	Internal standard (RT)	Deriv.	Other compounds (RT)	Ref.
B (3)	1	1.2 × 2	2% OV-101 Chromosorb® W (100/120)	275	Helium (30)	NPD	NA[a]	Strychnine (NA)	—	—	1
B (1)	1	1 × 3	3% OV-17 Gas Chrom® Q (100/120)	290	Nitrogen (60)	ECD	7.5	Papaveraldine (10.5)	—	—	2
U (20)	2	1.8 × 2	3% OV-17 Gas Chrom® Q (60/80)	NA	Nitrogen (30)	FID	NA	6′-Bromopapaverine (18.1)	—	Metabolite A[b] (9.8) Metabolite B[c] (13.9) Metabolite C[d] (8.9) Metabolite D[e] (8.2)	3

[a] Retention times of papaverine (2.3 min) and strychnine (5 min) are given using a 1.2 m × 3 mm column packed with 3% OV-17 and an FID detector.
[b] 1-(3-Methoxy-4-hydroxybenzyl)-6-7-dimethoxyisoquinoline.
[c] 1-(3,4-Dimethoxybenzyl)-6-methoxy-7-hydroxyisoquinoline.
[d] 1-(3,4-Dimethoxybenzyl)-6-hydroxy-7-methoxyisoquinoline.
[e] 1-(3-Methoxy-4-hydroxybenzyl)-6-hydroxy-7-methoxyisoquinoline.

REFERENCES

1. **Bellia, V., Jacob, J., and Smith, H. T.,** *J. Chromatogr.*, 161, 231, 1978.
2. **Zuccato, E., Marcucci, F., and Mussini, E.,** *J. Pharmacol. Methods*, 1, 9, 1978.
3. **Belpaire, F. M., Rosseel, M. T., and Bogaert, M. G.,** *Xenobiotica*, 8, 297, 1978.

PARALDEHYDE

Gas Chromatography

Specimen (mℓ)	S	Column m × mm	Packing (mesh)	Oven temp (°C)	Gas (mℓ/min)	Det.	RT (min)	Internal standard (RT)	Deriv.	Other compounds (RT)	Ref.
B (0.1)	3	1.8 × 2 Steel	0.4% Carbowax® Carbopack® A (NA)	130	NA (35)	FID	1.1[a]	*t*-Butanol (3.1)	—	Ethanol (1.4) Acetone (1.9) Isopropanol (2.1)	1

[a] Retention time of acetaldehyde formed by hydrolysis of paraldehyde with 10% sulfuric acid. Automatic sampling of head space.

REFERENCE

1. Hancock, J. P., Harrill, J. C., and Solomons, E. T., *J. Anal. Toxicol.*, 1, 161, 1977.

PARGYLINE

Gas Chromatography

Specimen (mℓ)	S	Column m × mm	Packing (mesh)	Oven temp (°C)	Gas (mℓ/min)	Det.	RT (min)	Internal standard (RT)	Deriv.	Other compounds (RT)	Ref.
B, U (10)	1	3 × 2 Steel	5% Carbowax® 6000 Chromosorb® G (80/100)	160	Nitrogen (20)	FID	16	*N,N*-Dimethylanline (6)	—	*N*-Methylbenzylamine (7)	1

Thin-Layer Chromatography

Specimen (mℓ)	S	Plate (Manufacturer)	Layer (mm)	Solvent	Post-separation treatment	Det. (nm)	R_f	Internal standard (R_f)	Other compounds (R_f)	Ref.
B, U (10)	3	20 × 20 cm Plastic sheet (Merck)	Silica gel 60 F_{254} (0.2)	S-1	—	Visual[a] (254)	NA	—	*N*-Methylbenzylamine (0.5)	1

Note: S-1 = Methanol + acetone + ammonia.
47.5 : 47.5 : 5

[a] The spots are scraped to measure radioactivity.

REFERENCE

1. Pirisino, R., Ciottoli, G. B., Buffoni, F., Anselmi, B., and Curradi, C., *Br. J. Clin. Pharmacol.*, 7, 595, 1979.

PEMOLINE

Gas Chromatography

Specimen (mℓ)	S	Column m × mm	Packing (mesh)	Oven temp (°C)	Gas (mℓ/min)	Det.	RT (min)	Internal standard (RT)	Deriv.	Other compounds (RT)	Ref.
U (10) B (2)	1	1.8 × 2	2% OV-17 Gas Chrom® Q (100/120)	160	Nitogen (30)	FID	16	Amobarbital (12.8)	i. Hydrolysis[a] ii. Methyl[b]	—	1
B (0.3)	1	0.9 × 2	3% Poly A-103 Gas Chrom® Q (100/120)	230	Nitrogen (45)	ECD	6	Methylpemoline (9)	Hydrolysis[a]	—	2
B (3) U (2)	1	8 × 0.5[c]	1.6% PPE-21 + 2.6% OV-17	190	Helium (5)	NPD	1.5	Fensuximide (2)	i. Hydrolysis[a] ii. Methyl[b]	—	3
Saliva			Cab-O-Sil (NA)								
B, U (1)	1	1.2 × 2	5% FFAP Chromosorb® W (80/100)	245	Helium (38)	NPD	4.2	2-Amino-5-(2-methylphenyl)-2-oxazolin-4-one (5)	Methyl[d]	—	4

High-Pressure Liquid Chromatography

Specimen (mℓ)	S	Column cm × mm	Packing (μm)	Elution	Flow rate (mℓ/min)	Det. (nm)	RT (min)	Internal standard (RT)	Other compounds (RT)	Ref.
U (10)	2	25 × 2.1	Zorbax® (NA)	E-1	0.5	UV (254)	5.5	—	—	5

PEMOLINE (continued)

Thin-Layer Chromatography

Specimen (mℓ)	S	Plate (manufacturer)	Layer (mm)	Solvent	Post-separation treatment	Det. (nm)	R_f	Internal standard (R_f)	Other compounds (R_f)	Ref.
U (NA)	3	20 × 20 cm (Laboratory)	Silica gel G (NA)	S-1 (Bidimensional)	Sp: i. 20% KOH in methanol ii. 1% *m*-Dinitrobenzene in methanol	Visual (red violet spot → orange spot)	i. 0.38[e] ii. 0.60	—	—	6

Note: E-1 = *n*-Hexane + 2-propanol + ammonia.
37 : 12 : 1

S-1 = i. Chloroform + methanol + 25% ammonia.
90 : 10 : 0.5
ii. *N N*-Dimethylformamide + ethayl acetate + *n*-octanol.
1 : 9 : 3 drops

[a] Pemoline is hydrolyzed with acid to 5-phenyl-2,4-oxazolidinedione.
[b] With diazomethane.
[c] Support-coated, open-tubular column.
[d] Extractive alkylation with methyl iodide in the presence of tetrapentylammonium hydroxide.
[e] R_f values in the 1st and 2nd dimensions.

REFERENCES

1. **van Boven, M. and Daenens, P.,** *J. Chromatogr.,* 134, 415, 1977.
2. **Chu, S. -Y. and Sennello, L. T.,** *J. Chromatogr.,* 137, 343, 1977.
3. **Vermeulen, N. P. E., Teunissen, M. W. E Breimer, D. D.,** *J. Chromatogr.,* 157, 133, 1978.
4. **Hoffman, D. J.,** *J. Pharm. Sci.,* 68, 445, 1979.
5. **Cartoni, G. P. and Natalizia, F.,** *J. Chromatogr.,* 123, 474, 1976.
6. **Goenechea, S. and Wagner, G. M.,** *J. Chromatogr.,* 140, 134, 1977.

PENBUTOLOL

Gas Chromatography

Specimen (mℓ)	S	Column m × mm	Packing (mesh)	Oven temp (°C)	Gas (mℓ/min)	Det.	RT (min)	Internal standard (RT)	Deriv.	Other compounds (RT)	Ref.
B (5)	2	3 × 3	(NA) OV-1 + 1% KOH Supelcoport (100/120)	200	Nitrogen (35)	FID	NA	Pronethalol (NA)	Trimethylsilyl	—	1

REFERENCE

1. Giudicelli, J. F., Richer, C., Chauvin, M., Idrissi, N., and Berdeaux, A., *Brit. J. Clin. Pharmacol.*, 4, 135, 1977.

PENICILLAMINE

High-Pressure Liquid Chromtography

Specimen (mℓ)	S	Column cm × mm	Packing (μm)	Elution	Flow rate (mℓ/min)	Det. (nm)	RT (min)	Internal standard (RT)	Other compounds (RT)	Ref.
B (0.8) U (2.25)	2	30 × 2	Zipax® SCX (NA)	E-1	0.6	Electro-chemical[a]	3	—	—	1

Note: E-1 = 0.04 *M* Phosphate/citrate buffer, pH 3.0.

[a] The oxidized penicillamine is reduced electrolytically prior to chromatography.

REFERENCE

1. **Saetre, R. and Rabenstein, D. L.,** *Anal. Chem.*, 50, 276, 1978.

PENICILLINS

1. Amoxicillin; 2. Ampicillin; 3. Penicillin G; 4. Penicillin V

Gas Chromatography

Specimen (mℓ)	S	Column m × mm	Packing (mesh)	Oven temp (°C)	Gas (mℓ/min)	Det.	RT (min)	Internal standard (RT)	Deriv.	Other compounds (RT)	Ref.
D	3	1.9 × 2	3% OV-17 Gas Chrom® Q (100/120)	60	Helum (30)	FID	—	Naphthalene (7)	—	*N,N*-Dimethylaniline[a] (15)	1
D	3	1 × 3	1.5% OV-17 Chromosorb® W (60/80)	Temp. Progr.	Helium (115)	FID	2^b = 16	5-α-Cholestanel (9)	Trimethylsilyl	α-Aminobenzylpenicilloic acid (12)	2

High-Pressure Liquid Chromatography

Specimen (mℓ)	S	Column cm × mm	Packing (μm)	Elution	Flow rate (mℓ/min)	Det. (nm)	RT (min)	Internal standard (RT)	Other compounds (RT)	Ref.
D	3	61 × 2.3	Bondapak® AX/ Corasil (37—50)	E-1	0.7	UV (254)	3^b = 17.5	—	Penicillamine (4.5) Benzylpenniloic acid (7.8) Benzylpenamaldic acid (13) Benzylpennilic acid (22)	3
D	3	100 × 2.1	Vydac® P150, AX (NA)	E-2	0.45	UV (254)	2^b = 6.5 3^b = 9 4^b = 16	—	—	4
B (0.1) Saliva (0.1) U (0.01)	1	15 × 4.6	LiChrosorb® RP-8 (5)	E-3	1.2	UV (225)	1^b = 6.5	—	c	5

PENICILLINS (continued)

1. Amoxicillin; 2. Ampicillin; 3. Penicillin G; 4. Penicillin V

High-Pressure Liquid Chromatography

Specimen (mℓ)	S	Column cm × mm	Packing (μm)	Elution	Flow rate (mℓ/min)	Det. (nm)	RT (min)	Internal standard (RT)	Other compounds (RT)	Ref.
U (5)	1	30 × 4	μ-Bondapak® C_{18} (10)	E-4	1.0	Fl[d] (Ex: 395, Fl 485)	1[b] = 7	—	Penicilloic acid of amoxicillin (5.2)	6
D	3	25 × 4.6	LiChrosorb® RP-8 (10)	E-5	1.0	UV (274)	4[b] = 12.3	1,3,5-Trimethoxybenzene	Benzathine (6) Penicillin V-penicilloic acid 6.4) Penicillin V-peniloic acid (7.2) Sodium benzoate (7.7) Methylparaben (8.1) Penicillin G (12.3) Propylparaben (18.2)	7
D	3	51 × 2.1	Partisil®-ODS (10—25)	E-6	1.0	UV (254)	3[b] = 7	—	Procaine (23)[e]	8
B (1)	2	10 × 4	LiChrosorb RP-8 (5)	E-7	1.0	UV[f] (310)	2[b] = 12	—	Mecillinam (17)	9

Thin-Layer Chromatgraphy

Specimen (mℓ)	S	Plate (manufacturer)	Layer (mm)	Solvent	Post-separation treatment	Det. (nm)	R_f	Internal standard (R_f)	Other compounds (R_f)	Ref.
D	3	20 × 20 cm (Laboratory)	Silica gel G (0.5)	S-1	Sp: i. 2*N* NaOH ii. Iodine-azide reagent iii. 1% Starch	Visual	3[b] = 0.66 4[b] = 0.66	—	Penillic acid (0.15)	10
D	3	20 × 20 cm (Merck)	Silica gel (NA)	S-2	—	UV Reflectance (230)	NA	—	—	11

Note: E-1 = 0.1 *M* Citric acid/0.2 *M* disodium phosphate buffer, pH 2.7.

E-2 = 0.02 *M* Sodium nitrate in 0.01 *M* sodium borate, pH 9.15.

E-3 = 0.067 *M* Potassium dihydrogen phosphate buffer, pH 4.6.

E-4 = Water + methanol + acetic acid.
85 : 15 : 0.5

E-5 = Methanol + 0.05 *M* sodium dihydrogen phosphate buffer, pH 3.5.
53 : 47

E-6 = Acetonitrile + water + 0.2 *M* ammonium acetate buffer, pH 6.
20 : 70 : 10

E-7 = Methanol + phosphate buffer, pH 8.
3 : 7

S-1 = Acetone + acetic acid.
95 : 5

S-2 = Acetone + chloroform + acetic acid.
50 : 45 : 5

[a] A contaminant in commercial penicillins.
[b] Numbers refer to a penicillin. See the top of the table.
[c] Conditions for the analysis of ampicillin in biological fluids are also given.
[d] Post-column reaction with fluorescamine.
[e] Degradation products of Penicillin G exposed to ^{60}Co irradiation are separated.
[f] Post-column reaction with 33% imidazole and 0.11% mercuric chloride at pH 7.2 prior to detection.

PENICILLINS (continued)

REFERENCES

1. Margosis, M., *J. Pharm. Sci.*, 66, 1634, 1977.
2. Wu, H. L., Msada, M., and Uno, T., *J. Chromatogr.*, 137, 127, 1977.
3. Blaha, J. M., Knevel, A. M., and Hem, S. L., *J. Pharm. Sci.*,64, 1384, 1975.
4. Tsuji, K. and Robertson, J. H., *J. Pharm. Sci.*, 64, 1542, 1975.
5. Vree, T. B., Hekster, Y. A., Baars, A. M., and van der Kleijn, E., *J. Chromatogr.*, 145, 496, 1978.
6. Lee, T. L., D'Arconte, L., and Brooks, M. A., *J. Pharm. Sci.*, 68, 454, 1979.
7. LeBelle, M., Graham, K. and Wilson, W. L., *J. Pharm. Sci.*, 68, 555, 1979.
8. Tsuji, K., Goetz, J. F., and Vanmeter, W., *J. Pharm. Sci.*, 68, 1075, 1979
9. Westerlund, D., Carlqvist, J., and Theodorsen, A., *Acta Pharm. Suec.*, 16, 187, 1979.
10. Vandamme, E. J. and Voets, J. P., *J. Chromatogr.*, 71, 141, 1972.
11. Manni, P. E., Bourgeois, M. F., Lipper, R. A., Blaha, J. M., and Hem. S. L., *J. Chromatogr.*, 85, 177, 1973.

PENTAMETHYLMELAMINE

Gas Chromatography

Specimen (mℓ)	S	Column m × mm	Packing (mesh)	Oven temp (°C)	Gas (mℓ/min)	Det.	RT (min)	Internal standard (RT)	Deriv.	Other compounds (RT)	Ref.
B, U (1)	1	0.6 × 2	10% Carbowax® 20 *M* + 2% KOH Chromosorb® W (80/100)	Temp. Progr.	Nitrogen (30)	NPD	4.2	Hexamethyl melamine (2.5)	—	—	1

REFERENCE

1. Ames, M. M. and Powis, G., *J. Chromatogr.*, 174, 245, 1979.

PENTAZOCINE

Gas Chromatography

Specimen (mℓ)	S	Column m × mm	Packing (mesh)	Oven temp (°C)	Gas (mℓ/min)	Det.	RT (min)	Internal standard (RT)	Deriv.	Other compounds (RT)	Ref.
B (0.5—2)	2	1.8 × 2	1% OV-1 Gas Chrom Q (100/120)	245	Nitrogen (55)	FID	NA	a (NA)	Trimethyl-silyl	Metabolites[b]	1
B (0.5)	2	0.9 × 2	5% OV-17 Gas Chrom® Q (80/100)	235	Nitrogen (30)	ECD	8	Ketohemidone (3.5)	Pentafluo-robenzyl	—	2
B (1)	2	1.5 × 4	5% QF-1 Supasorb (60/80)	206	Argon (30)	NPD	13.9	Methylhexyl-mercapturate (10.1)	—	—	3
B, U (0.2—1)	2	1.8 × 3.2	3% Dexsil® — 300 Gas Chrom® Q (100/120)	265	Argon 95 + Methane 5 (50)	ECD	4.8	Levallorphan (6.2)	Pentafluo-robenzyl	—	4

Thin-Layer Chromatography

Specimen (mℓ)	S	Plate (manufacturer)	Layer (mm)	Solvent	Post-separation treatment	Det. (nm)	R_f	Internal standard (R_f)	Other compounds (R_f)	Ref.
U (20—25)	3	20 × 20 cm (Gelman)	Silica gel -SA (NA)	S-1	Sp: i. Iodoplatinate reagent ii. Iodine-potassium io-dide reagent	Visual	0.35[c]	—	Phencyclidine (0.82) Propoxyphene (0.71) Diphenhydramine (0.4)	5

Note: S-1 = Ethyl acetate + cyclohexane + methanol + ammonia.
56 : 40 : 0.8 : 0.4

[a] 1,2,3,4,5,6-Hexahydro-cis-6,11-dimethyl-3-cyclopentyl-2,6-methano-3-benzazocin-8-ol (an analog of pentazocine).
[b] Analysis of urinary metabolites is described. However, chromatograms are not shown.
[c] A range of r_f values is given.

REFERENCES

1. **Pittman, K. A. and Davison, C.**, *J. Pharm. Sci.*, 62, 765, 1973.
2. **Brötell, H., Ehrsson, H., and Gyllenhall, P.**, *J. Chromatogr.*, 78, 293, 1973.
3. **James, S. P. and Waring, R. H.**, *J. Chromatogr.*, 78, 417, 1973.
4. **Swezey, S. E., Blaschke, T. F., and Meffin, P. J.**, *J. Chromatogr.*, 154, 256, 1978.
5. **Kaistha, K. K. and Tadrus, R.**, *J. Chromatogr.*, 155, 214, 1978.

PENTOBARBITAL

Gas Chromatography

Specimen (mℓ)	S	Column m × mm	Packing (mesh)	Oven temp (°C)	Gas (mℓ/min)	Det.	RT min)	Internal standard (RT)	Deriv.	Other compounds (RT)	Ref.
B (0.1)	1	1.2 × 2	3% OV-101 Chromosorb® (100/120)	140	Helium (40)	NPD	3.1	Secobarbital (3.8)	Methyl[a]	b	1[c]

[a] With methyl iodide in acetone in the presence of sodium carbonate.
[b] Other barbiturates can also be analyzed by this procedure.
[c] The authors claim that their earlier procedure for the analysis of pentobarbital by gas chromatography with the use of an electron-capture detector (*J. Pharm. Sci.*, 66, 477, 1977) is tedious because of numerous clean-up steps for sample preparation.

REFERENCE

1. **Sun, S. -R. and Hoffman, D. J.**, *J. Pharm. Sci.*, 68,386, 1979.

PENTYLENETETRAZOLE

Gas Chromatography

Specimen (mℓ)	S	Column m × mm	Packing (mesh)	Oven temp (°C)	Gas (mℓ/min)	Det.	RT (min)	Internal standard (RT)	Deriv.	Other compounds (RT)	Ref.
B (0.5)	2	1.8 × 3 Aluminum	5% Carbowax® 20 *M* Chromosorb® W (80/100)	200	Helium (65)	FID	2.5	Procaine (9)	—	—	1
B (1.5)	1	1 × 4	3% OV-17 Gas Chrom® Q (100/120)	162	Nitrogen (30)	NPD	5	Iproniazid (3)	—	—	2

REFERENCES

1. Stewart, J. T. and Story, J. L., *J. Pharm. Sci.*, 61, 1651, 1972.
2. Dal Bo, L., Marcucci, F., and Mussini, E., *J. Pharmacol. Methods*, 2, 29, 1979.

PERAZINE

Gas Chromatography

Specimen (mℓ)	S	Column m × mm	Packing (mesh)	Oven temp (°C)	Gas (mℓ/min)	Det.	RT: (min)	Internal standard (RT)	Deriv.	Other compounds (RT)	Ref.
B (5)	2	1.8 × 4	3% XE-60® Gas Chrom Q (80/100)	240	Nitrogen (30)	FID	3.3	Thioridazine (8)	—	a	1

[a] Conditions are described for the assay of thioridazine and metabolites using perazine as the internal standard.

REFERENCE

1. Vanderheeren, F. A. J., Theunis, D. J. C. J., and Rosseel, M.-T., *J. Chromatogr.*, 120, 123, 1976.

PERPHENAZINE

Gas Chromatograph

Specimen (mℓ)	S	Column m × mm	Packing (mesh)	Oven temp (°C)	Gas (mℓ/min)	Det.	RT (min)	Internal standard (RT)	Deriv.	Other compounds (RT)	Ref.
B (2.5)	1	1.5 × 4	1% OV-17 Celite® JJCQ (100/120)	305	Argon 90 + Methane 10 (60)	ECD	NA	8-Chloroperphenazine (NA)	Trimethylsilyl	Perphenazinesulfoxide (NA)	1
B (3)	1	2 × 4	1% OV-17® Chromosorb W (100/120)	280	Argon 95 + Methane 5 (50)	ECD	11.3	Thioproperazine (26.9)	Trimethylsilyl	a	2

[a] Conditions for the separate estimation of amitriptyline and nortriptyline are also described.

REFERENCES

1. Larsen, N.-E. and Naestoft, J., *J. Chromatgr.*, 109, 259, 1975.
2. Cooper, S., Albert, J. M., Dugal, R., Bertrand, M., and Elie, R., *Arzneim Forsch.*, 29(I), 158, 1979; *J. Chromatogr.*, 150, 263, 1978.

PHANQUONE

Gas Chromatography

Specimen (mℓ)	S	Column m × mm	Packing (mesh)	Oven temp (°C)	Gas (mℓ/min)	Det.	RT (min)	Internal standard (RT)	Deriv.	Other compounds (RT)	Ref.
B, U (1—3)	1	1.5 × 2	3% JXR Chromosorb® G (80/100)	210	Nitrogen (30)	ECD	10	10-Methyl-4,7-phenanthro-line-5,6-dione (13.7)	Methoxine	—	1

REFERENCE

1. Degen, P. H., Brechbühler, S., Schäublin, J., and Riess, W., *J. Chromatogr.*, 118, 363, 1976.

PHENACETIN

High-Pressure Liquid Chromatography

Specimen (mℓ)	S	Column cm × mm	Packing (μm)	Elution	Flow rate (mℓ/min)	Det. (nm)	RT (min)	Internal standard (RT)	Other compounds (RT)	Ref.
B (0.5—2) U (0.005—0.1)	2	30 × 4	μ-Bondapak® C_{18} (10)	E-1	2	UV (254)	6.4	—	—	1
B (0.1)	2	30 × 4	μ-Bondapak® C_{18} (10)	E-2	1	UV (254, 280)	9	^{14}C-Phenacetin (9)	Acetaminophen (4)	2

Note: E-1 = Methanol + 0.01 *M* $(NH_4)_2CO_3$.
3 : 7

E-2 = Acetonitrile + water + PIC B-7.
30 : 70 : 1.6

REFERENCES

1. **Duggin, G. G.,** *J. Chromatogr.,* 121, 156, 1976.
2. **Pang, K. S., Taburet, A. M., Hinson, J. A., and Gillete, J. R.,** *J. Chromatogr.,* 174, 165, 1979.

PHENCYCLIDINE

Gas Chromatography

Specimen (mℓ)	S	Column m × mm	Packing (mesh)	Oven temp (°C)	Gas (mℓ/min)	Det.	RT (min)	Internal standard (RT)	Deriv.	Other compounds (RT)	Ref.
B (0.25)	1	1.2 × NA	3% OV-17 Gas Chrom® Q (100/120)	190	Helium (40)	MS-EI	NA	[$^2H_5$] Phencyclidine (NA)	—	—	1
B (NA)	1	1.5 × 2	3% OV-17 Gas Chrom® Q (NA)	190	Helium (20)	MS-CI	2.8	[$^2H_5$] Phencyclidine (2.8)	—	—	2
B, U (5)	3	1.2 × NA	3% OV-17 Gas Chrom® Q (100/120)	155	NA	FID	4.2	*n*-Docosane (7)	—	—	3[a]
B (1)	2	1.2 × NA	3% OV-1 Chromosorb® W (80/100)	185	Helium (30)	NPD	3.5	Benzphetamine (2.5)	—	—	4
D	3	1.2 × 2	3.8% UC W-98 Gas Chrom® Q (80/100)		Nitrogen (55)	FID	3.2	*N*-Methyl-*N*-propyl-1-phenylcylohexylamine (1.6)	—	1-Piperidinocyclohexanecarbonitrile[b] (0.9) 1-Piperidinocyclohexene (0.4)	5

Thin-Layer Chromatography

Specimen (mℓ)	S	Plate (manufacturer)	Layer (mm)	Solvent	Post-separation treatment	Det. (nm)	R_f	Internal standard (R_f)	Other compounds (R_f)	Ref.
U (20)	3	NA (Merck)	Silica gel G (NA)	S-1[c]	Sp: Acidified iodoplatinate reagent	Visual	0.58	—	Benzoylecgonine (0.03) Codeine (0.20) Propoxyphene (0.59)	6

U (5)	3	NA (Analytical Systems)[d]	Silica (NA)	S-2	Sp: Modified Dragendorf reagent	Visual	0.9[e]	—	—	7

Note: S-1 = Ethyl acetate + methanol + diethylamine.
90 : 10 : 1.6

S-2 = Ethyl acetate + metahnol + ammonia + water.
29 : 1 : 0.25 : 0.5

[a] Case histories of fatal poisoning due to ingestion of phencyclidine are given.
[b] Alternative conditions for the assay of the compound are described.
[c] An alternative solvent is also described.
[d] This company markets a complete "Toxi-Lab" kit for screening of drugs in urine.
[e] Some drugs with similar R_f values are eliminated by sequential spraying with different reagents. Phencyclidine is also confirmed by developing the plate in an alternative solvent.

REFERENCES

1. **Pearce, D. S.,** *Clin. Chem.*, 22, 1623, 1976.
2. **Wilson, A. E. and Domino, E. F.,** *Biomed. Mass Spectrom.*, 5, 112, 1978.
3. **Caplan, Y. H., Orloff, K. G., Thompson, B. C., and Fisher, R. S.,** *J. Anal. Toxicol.*, 3, 47, 1979.
4. **Lewellen, L. J., Solomons, E. T., and O'Brien, F. L.,** *J. Anal. Toxicol.*, 3, 72, 1979.
5. **Ballinger, J. R. and Marshman, J. A.,** *J. Anal. Toxicol.*, 3, 158, 1979.
6. **Jain, N. C., Budd, R. D., Leung, W. J., and Sneath, T. C.,** *J. Anal. Toxicol.*, 1, 77, 1977.
7. **Finkle, H. I.,** *Am. J. Clin. Pathol.*, 70, 287, 1978.

PHENDIMETRAZINE

Gas Chromatography

Specimen (mℓ)	S	Column m × mm	Packing (mesh)	Oven temp (°C)	Gas (mℓ/min)	Det.	RT (min)	Internal standard (RT)	Deriv.	Other compounds (RT)	Ref.
B (5)	1	1.8×NA	3% SE-30 Chromosorb® W (80/100)	Temp. Progr.	Nitrogen (80)	FID	4.9	Diethylpropion (5.8)	—	—	1

REFERENCE

1. Hundt, H. K. L., Clark, E. C., and Müller, F. O., *J. Pharm. Sci.*, 64, 1041, 1975.

PHENFORMIN

Gas Chromatography

Specimen (mℓ)	S	Column m × mm	Packing (mesh)	Oven temp (°C)	Gas (mℓ/min)	Det.	RT (min)	Internal standard (RT)	Deriv.	Other compounds (RT)	Ref.
B (1)	1	1.8 × 3	3% OV-17 Chromosorb® W (100/120)	210	Argon 95 + Methane 5 (30)	ECD	5.8	Benzylbiguanide (4.2)	Monochlorodifluoroacetyl	Buformin[a] Metformin[a]	1
B (0.2)	1	1.8 × 2 Steel	10% SE-30 Chromosorb® W (80/100)	208	Helium (25)	FID	4.8	Naphthylamine (1)	Trifluoroacetyl	*p*-Hydroxyphenformin (7.8)	2

High-Pressure Liquid Chromatography

Specimen (mℓ)	S	Column cm × mm	Packing (μm)	Elution	Flow rate (mℓ/min)	Det. (nm)	RT (min)	Internal standard (RT)	Other compounds (RT)	Ref.
B (1)	1	30 × 4	μ-Bondapak® C_{18} (10)	E-1	1.0	UV (233)	6	—	Tolbutamide[b]	3

Note: E-1 = Methanol + 0.02% acetic acid containing 0.005 *M* 1-heptanesulfonic acid, pH 3.7.
1 : 1

[a] Conditions for the analysis of these compounds are described.
[b] Tolbutamide is analyzed using a different mobile phase.

REFERENCES

1. Matin, S. B., Karam, J. H., and Forsham, P. H., *Anal. Chem.*, 47, 545, 1975.
2. Mottale, M. and Stewart, C. J., *J. Chromatogr.*, 106, 263, 1975.
3. Hill, H. M. and Chamberlain, J. *J. Chromatogr.*, 149, 349, 1978.

PHENOBARBITAL

Gas Chromatography

Specimen (mℓ)	S	Column m × mm	Packing (mesh)	Oven temp (°C)	Gas (mℓ/min)	Det.	RT (min)	Internal standard (RT)	Deriv.	Other compounds (RT)	Ref.
B(0.25)	2	1.8 × 4	3% XE-60 Gas Chrom® Q (100/120)	190	Nitrogen (30)	FID	5.9	5-Ethyl-5-*p*-tolylbarbituric acid (7.6)	Methyl[a]	—	1
B (1)	2	1.8 × 2 steel	10% UC-W98 Chromosorb® W (80/100)	200	Nitrogen (35)	FID	6	Cyclobarbital (NA)	Methyl[b]	—	2
B (0.05)	2	1.5 × 4	2% SP-525[c] Varaport 30 (80/100)	260	Argon (60)	FID	1.6	Tetraphenylethylene (2.9)	—	Primidone (5.4) Diphenylhydantoin (6.7)	3

High-Pressure Liquid Chromatography

Specimen (mℓ)	S	Column cm × mm	Packing (μm)	Elution	Flow rate (mℓ/min)	Det. (nm)	RT (min)	Internal standard (RT)	Other compounds (RT)	Ref.
Animal chow	3	30 × 4	μ-Bondapak® C_{18} (10)	E-1	1.0	UV (210)	7	—	—	4

Thin-Layer Chromatography

Specimen (mℓ)	S	Plate (manufacturer)	Layer (mm)	Solvent	Post-separation treatment	Det. (nm)	R_f	Internal standard (R_f)	Other compounds (R_f)	Ref.
U (5)	2	20 × 20 cm (Analtech)	Silica gel GF (0.25)	S-1	—	Quenching of fluorescence (255)	0.68	—	*p*-Hydroxypheno-barbital (0.79)	5

[a] The dried extracts are heated with 0.2 *M* trimethylanilinium hydroxide at 100° for 5 min prior to gas chromatography. This technique minimizes the decomposition of methylated phenobrbital in the presence of derivatizing agent.

[b] On-column methylation with 2 *M* trimethylanilinium hydroxide in the presence of 50% glycerol in methanol to suppress the decomposition of methylated phenobarbital.

[c] The column was conditioned by periodic injection of 10 to 20 $\mu\ell$ of γ-glycidoxypropyltrimethoxysilane (H-187).

Note: E-1 = Water + methanol.
60 : 40

S-1 = Ethyl acetate + methanol + ammonia.
82 : 15 : 4

REFERENCES

1. Gupta, R. N. and Keane, P. M., *Clin. Chem.*, 21, 1346, 1975.
2. Kelly, R. C., Valentour, J. C., and Sunshine, I., *J. Chromatogr.*, 138, 413, 1977.
3. Rutherford, D. M. and Flanagen, R. J., *J. Chromatogr.*, 157, 311, 1978.
4. Bowman, M. C. and Rushing, L. G., *J. Chromatogr. Sci.*, 16, 23, 1978.
5. Levin, S. S., Schwartz, M. F., Cooper, D. Y., and Touchstone, J. C., *J. Chromatogr.*, 154, 349, 1978; *J. Chromatogr. Sci.*, 15, 528, 1977.

PHENOL

High-Pressure Liquid Chromatography

Specimen (mℓ)	S	Column cm × mm	Packing (μm)	Elution	Flow rate (mℓ/min)	Det. (nm)	RT (min)	Internal standard (RT)	Other compounds (RT)	Ref.
D	3	30 × 4	μ-Bondapak® C_{18} (10)	E-1	3.0	UV (254)	6	—	Resorcinol (3)	1

Note: E-1 = Methanol + water + NH_4HCO_3, pH 7.67.
100 : 900 : (0.02 moles)

REFERENCE

1. Gupta, V. D., *J. Pharm. Sci.*, 65, 1706, 1976.

PHENPROCOUMON

Gas Chromatography

Specimen (mℓ)	S	Column m × mm	Packing (mesh)	Oven temp (°C)	Gas (mℓ/min)	Det.	RT (min)	Internal standard (RT)	Deriv.	Other compounds (RT)	Ref.
B (2)	1	1.8 × 3 Steel	5% OV-25 Chromosorb® W (80/100)	260	Nitrogen (60)	FID	9.4	Diphenylhydantoin (5.7)	Methyl[a]	—	1
B (2)	2	1.8 × 2	3% OV-225 Gas Chrom® Q (100/120)	250	Nitrogen (b)	FID	5.5 (4.5)[c]	—	Methyl[a]	—	2
B (2)	1	1.8 × 2	3% OV-17 Chromosorb® W (100/120)	265	Nitrogen (50)	FID	6.5	*p*-Chlorophenprocoumon (12)	—	—	3

High-Pressure Liquid Chromatography

Specimen (mℓ)	S	Column cm × mm	Packing (μm)	Elution	Flow rate (mℓ/min)	Det. (nm)	RT (min)	Internal standard (RT)	Other compounds (RT)	Ref.
B (1)	1	80 × 2	Merckosorb® SI60 (5)	E-1	0.4	UV (254)	4	4-Hydroxycoumarin (6)	—	4

Thin-Layer Chromatography

Specimen (mℓ)	S	Plate (manufacturer)	Layer (mm)	Solvent	Post-separation treatment	Det. (nm)	R_f	Internal standard (R_f)	Other compounds (R_f)	Ref.
B (1)	2	20 × 20 cm (Merck)	Silica gel (0.25)	S-1	—	Fl (Ex: 313, Fl: 410)	0.55	—	c	5
B (0.1—0.2)	1	20 × 20 (Merck)	Silica gel 60 F_{254} (0.25)	S-2	—	Fl (Ex: 312, Fl: 365)	0.45	—	5-Hydroxyphenprocoumon (0.23)	6

PHENPROCOUMON (continued)

Thin-Layer Chromatography

Specimen (mℓ)	S	Plate (manufacturer)	Layer (mm)	Solvent	Post-separation treatment	Det. (nm)	R_f	Internal standard (R_f)	Other compounds (R_f)	Ref.
									6-Hydroxyphenprocoumon (0.23) 6-Hydroxyphenprocoumon (0.14)	

Note: E-1 = Methanol.

S-1 = Chloroform + methanol.
97 : 3

S-2 = Chloroform + methanol + triethylamine.
95 : 15 : 5

[a] On-column methylation with trimethylanilinium hydroxide.
[b] Inlet pressure = 30 psi.
[c] A second minor peak.
[d] R_f values of a number of drugs are given to show their noninterference with the assay of phenprocoumon.

REFERENCES

1. Midha, K. K., Hubbard, J. W., Cooper, J. K., and McGilveray, I. J., *J. Pharm. Sci.*, 65, 387, 1976.
2. Brombacher, P. J., Cremers, H. M. H. G., Mol, M. J., Muijers, P. H. J., van der Plas, P. M., and Verheesen, P. E., *Clin. Chim. Acta*, 75, 443, 1977.
3. Schmitt, K.-F. and Jähnchen, E., *J. Chromatogr.*, 130,418, 1977.
4. Kinawi, A., *Arzneim. Forsch.*, 27(1), 360, 1977.
5. de Wolff, F. A. and van Kempen, G. M. J., *Clin. Chem.*, 22, 1575, 1976.
6. Haefelfinger, P., *J. Chromatogr.*, 162, 215, 1979.

PHENTERMINE

Gas Chromatography

Specimen (mℓ)	S	Column m × mm	Packing (mesh)	Oven temp (°C)	Gas (mℓ/min)	Det.	RT (min)	Internal standard (RT)	Deriv.	Other compounds (RT)	Ref.
B (30)	1	2×NA Steel	10% Apiezon® L + 10% KOH Chromosorb® G (80/100)	200	NA	NPD	4.5	Dimethylamphetamine (6)	—	—	1

REFERENCE

1. Vycudilik, W., *J. Chromatogr.*, 111, 439, 1975.

1-(2-PHENYLADMANT-1-YL)-2-METHYLAMINOPROPANE

Gas Chromatography

Specimen (mℓ)	S	Column m × mm	Packing (mesh)	Oven temp (°C)	Gas (mℓ/min)	Det.	RT (min)	Internal standard (RT)	Deriv.	Other compounds (RT)	Ref.
B (2—3)	1	1.5×NA	5% XE-60 Gas Chrom® Q (60/85)	190	Nitrogen (80)	ECD	I[a] = 8.5 II = 10	—	Chlorodi-fluoro ace-tyl	—	1

[a] Diastereoisomers.

REFERENCE

1. Cockerill, A. F., Mallen, D. N. B., Osborne, D. J., and Prime, D. M., *J. Chromatogr.*, 114, 151, 1975.

PHENYLBUTAZONE

Gas Chromatography

Specimen (mℓ)	S	Column m × mm	Packing (mesh)	Oven temp (°C)	Gas (mℓ/min)	Det.	RT (min)	Internal standard (RT)	Deriv.	Other compounds (RT)	Ref.
B (1)	1	0.9×NA	3% Apiezon® L Chromosorb® W (80/100)	230	Ntrogen (120)	FID	3.5	Diphenylphthalate(6.0)	—	—	1
B (0.1)	1	NA	5% OV-17 Chromosorb® W (80/100)	250	Argon 90 + Methane 10 (70)	ECD	NA	4-Butyl-1,2-bis (*p*-tolyl)-3,5-pyrazolidine (NA)	—	—	2
B, U (5)	2	1.8 × 2	3% SE-30 Supelcoport (80/100)	220	Nitrogen (45)	ECD	4	—	Trimethylsilyl	Oxyphenbutazone (12)	3

High-Pressure Liquid Chromatography

Specimen (mℓ)	S	Column cm × mm	Packing (μm)	Elution	Flow rate (mℓ/min)	Det. (nm)	RT (min)	Internal standard (RT)	Other compounds (RT)	Ref.
B (1)	1	100 × 1.8	Sil-X[a] (NA)	E-1	1	UV (254)	2.5	2,4-Dinitrophenylhydrazone of 3,4-dimethoxybenzaldehyde (4.5)	Oxyphenbutazone (6)	4

Note: E-1 = *n*-Hexane + tetrahydrofuran + acetic acid.
77 : 23 : 0.002

[a] Column temperature = 37°C.

PHENYLBUTAZONE (continued)

REFERENCES

1. McGilveray, I. J., Midha, K. K., Brien, R., and Wilson, L., *J. Chromatogr.*, 89, 17, 1974.
2. Sioufi, A., Caydal, F., and Marfil, F., *J. Pharm. Sci.*, 67, 243, 1978.
3. Norheim, G., Hole, R., Froslie, A., and Bergsjo, T. H., *Fresenius Z. Anal. Chem.*, 289, 287, 1978.
4. Pound, N. J. and Sears, R. W., *J. Pharm. Sci.*, 64, 284, 1975.

PHENYLEPHRINE

Gas Chromatography

Specimen (mℓ)	S	Column m × mm	Packing (mesh)	Oven temp (°C)	Gas (mℓ/min)	Det.	RT (min)	Internal standard (RT)	Deriv.	Other compounds (RT)	Ref.
B (1—2)	2	1.8 × 2	3% QF-1 Gas Chrom® Q (100/120)	150	Helium (50)	ECD	3.8	—	Trifluoroacetyl	—	1

High-Pressure Liquid Chromatography

Specimen (mℓ)	S	Column cm × mm	Packing (μm)	Elution	Flow rate (mℓ/min)	Det. (nm)	RT (min)	Internal standard (RT)	Other compounds (RT)	Ref.
D	3	30 × 4	μ-Bondapak®-phenyl (10)	E-2	2	UV (254)	2	—	Pseudoephedrine (3) Naphazoline (6.5) Chlorpheniramine (11) Brompheniramine (13)	3
D	3	30 × 4	μ-Bondapak®-CN (10)	E-1	0.6[a]	UV (254)	8.5	—	Phenylpropanolamine (9) Benzoic acid (11) Brompheniramine (15.5)	2

Note: E-1 = Acetonitrile + 1.8% acetic acid + sodium heptanesulfonate.
130 : 870 : (0.005 moles)

E-2 = Water + methanol + acetic acid + sodium heptanesulfonate.
550 : 440 : 10 : (0.005 moles)

[a] This flow rate is changed to 3.6 mℓ/min after 12 min.

PHENYLEPHRINE (continued)

REFERENCES

1. Dombrowski, L. J., Comi, P. M., and Pratt, E. L., *J. Pharm. Sci.*, 62, 1761, 1973.
2. Ghanekar, A. G. and Gupta, V. D., *J. Pharm. Sci.*, 67, 873, 1978; ibid, 67,1247, 1978.
3. Koziol, T. R., Jacob, J. T., and Achari, R. G., *J. Pharm. Sci.*, 68, 1135, 1979.

PHENYLPROPANOLAMINE

Gas Chromatography

Specimen (mℓ)	S	Column m × mm	Packing (mesh)	Oven temp (°C)	Gas (mℓ/min)	Det.	RT (min)	Internal standard (RT)	Deriv.	Other compounds (RT)	Ref.
B (3)	1	2 × 2	1.25% OV-17 Gas Chrom® Q (100/120)	190	Nitrogen (30)	ECD	5.5	2,4-Dinitrophenyl-*N,N*-diethylamine (14)	a	—	1
D	3	1.8 × 4	3% OV-17 Gas Chrom® Q (100/120)	230	Helium (20)	FID	0.9	*n*-Tetrasocaine (5.8)	—	Chlorpheniramine (4.6)	2
U (1)	2	2.2 × 2 Steel	3% OV-1 Gas Chrom® Q (100/120)	230	Nitrogen (35)	NPD	0.3	Promethazine (3.4)	—	Chlorpheniramine (1.6)	3

[a] Plasma is treated with pentafluorobenzaldehyde to prepare pentafluorophenyloxazolidine derivative.

REFERENCES

1. Neelakantan, L. and Kostenbauder, H. B., *J. Pharm. Sci.*, 65, 740, 1976.
2. Madsen, R. E. and Magin, D. F., *J. Pharm. Sci.*, 65, 924, 1976.
3. Kinsun, H., Moulin, M. A., and Savini, E. C., *J. Pharm. Sci.*, 67, 118, 1978.

PILOCARPINE

High-Pressure Liquid Chromatography

Specimen (mℓ)	S	Column cm × mm	Packing (μm)	Elution	Flow rate (mℓ/min)	Det. (nm)	RT (min)	Internal standard (RT)	Other compounds (RT)	Ref.
D	3	10 × 6	Aminex® A-7 (7—11)	E-1	0.4	UV (217)	45	—	Isopilocarpine (38)	1

Note: E-1 = 2-propanol + 0.2 *M* tris buffer, pH 9.
5 : 95

REFERENCE

1. Urbanyi, T., Piedmont, A., Willis, E., and Manning, G., *J. Pharm. Sci.*, 65, 257, 1976.

PINAZEPAM

Gas Chromatography

Specimen (mℓ)	S	Column m × mm	Packing (mesh)	Oven temp (°C)	Gas (mℓ/min)	Det.	RT (min)	Internal standard (RT)	Deriv.	Other compounds (RT)	Ref.
B (0.5) U (2) Milk (1)	1	2.5 × 4	3% OV-17 Gas Chrom® Q (100/120)	264	Nitrogen (60)	ECD	5.7	Lorazepam (4.2)	—	Oxazepam (3,3) *N*-Demethyldiaze-pam (6.3) 3-Hydroxypinaze-pam (11.4)	1

High-Pressure Liquid Chromatography

Specimen (mℓ)	S	Column cm × mm	Packing (μm)	Elution	Flow rate (mℓ/min)	Det. (nm)	RT (min)	Internal standard (RT)	Other compounds (RT)	Ref.
B (0.5)	1	25 × NA	LiChrosorb® RP-8[a] (10)	E-1	0.8	UV (254)	14.6	Diazepam (11.7)	*N*-Desmethyldi-azepam (8.7) Oxazepam (7) 3-Hydroxypinaze-pam (10.4)	2

Note: E-1 = Water + acetonitrile
45 : 55

[a] Column temperature = 40°C.

REFERENCES

1. **Pacifici, G. M. and Placidi, G. F.,** *J. Chromatogr.*, 135, 133, 1977.
2. **Grassi, E., Passetti, G. L., and Trebbi, A.,** *J. Chromatogr.*, 144, 132, 1977.

PIPERAZINE

High-Pressure Liquid Chromatography

Specimen (mℓ)	S	Column cm × mm	Packing (μm)	Elution	Flow rate (mℓ/min)	Det. (nm)	RT (min)	Internal standard (RT)	Other compounds (RT)	Ref.
Tissue	2	61 × 2.1	Chroma-sep Silica (37)	E-1	1.2	Absorbance (363)	—	—	Mononitrosopiperazine (1.8) Dinitrosopiperazine (5.8)	1

Note: E-1 = Ethyl acetate + methanol.
1 : 1

REFERENCE

1. Rao, G. S. and McLennon, D. A., *J. Anal. Toxicol.*, 1, 43, 1977.

PIRACETAM

Gas Chromatography

Specimen (mℓ)	S	Column m × mm	Packing (mesh)	Oven temp (°C)	Gas (mℓ/min)	Det.	RT (min)	Internal standard (RT)	Deriv.	Other compounds (RT)	Ref.
B (0.5)	2	1 × 2	Tenax GC (60/80)	275	Helium (54)	NPD	2.2	Propionyl analog (4)	—	—	1

REFERENCE

1. Hesse, C. and Schulz, M., *Chromatographia*, 12, 12, 1979.

POLYTHIAZIDE

High-Pressure Liquid Chromatography

Specimen (mℓ)	S	Column cm × mm	Packing (μm)	Elution	Flow rate (mℓ/min)	Det. (nm)	RT (min)	Internal standard (RT)	Other compounds (RT)	Ref.
D	3	25 × 4.6	Partisil® 10-ODS (10)	E-1	0.4	UV (254)	7.0	Quinoline (10)	a Vanillin[b] (3.3)	1
D	3	25 × 4.6	Partisil® 10-ODS (10)	E-2	1.0	UV (254)	3.0	—	Vanillin (6.5)	2

Note: E-1 = Methanol + water.
35 : 65

E-1 = Acetonitrile + water.
980 : 20

[a] Chromatographic behavior of some thiazide and non-thiazide diuretics is described.
[b] Tablet excipient.

REFERENCES

1. Moskalyk, R. E., Locock, R. A., Chatten, L. G., Veltman, A. M., and Bielech, M. F., *J. Pharm. Sci.*, 64, 1406, 1975.
2. Wong, C. K., Tsau, D. Y. J., Cohen, D. M., and Munnelly, K. P., *J. Pharm. Sci.*, 66, 736, 1977.

PRACTOLOL

Gas Chromatography

Specimen (mℓ)	S	Column m × mm	Packing (mesh)	Oven temp (°C)	Gas (mℓ/min)	Det.	RT (min)	Internal standard (RT)	Deriv.	Other compounds (RT)	Ref.
B (1)	1	3.2 × 3	4% SE-30 Gas Chrom® Q (80/100)	200	Nitrogen (60)	ECD	4.6	Propranolol (6)	Trifluoacetyl	—	1

High-Pressure Liquid Chromatography

Specimen (mℓ)	S	Column cm × mm	Packing (μm)	Elution	Flow rate (mℓ/min)	Det. (nm)	RT (min)	Internal standard (RT)	Other compounds (RT)	Ref.
B (1)	2	30 × 4	μ-Bondapak® C_{18} (10)	E-1	1.0	UV (254)	8	—	—	2

Note: E-1 = 0.1% Phosphoric acid in ethanol + water.
1 : 9

REFERENCES

1. **Desager, J. P. and Harvengt, C.,** *J. Pharm. Pharmacol.*, 27, 52, 1975.
2. **Cooper, M. J. and Mirkin, B. L.,** *J. Chromatogr.*, 163, 244, 1979.

PRAZOSIN

High-Pressure Liquid Chromatography

Specimen (mℓ)	S	Column cm × mm	Packing (μm)	Elution	Flow rate (mℓ/min)	Det. (nm)	RT (min)	Internal standard (RT)	Other compounds (RT)	Ref.
B (2)	2	30 × 4	μ-Bondapak®-CN (10)	E-1	2.4	Fl (Ex: 246, Fl: 389)	NA	a (NA)	—	1
B (0.1—1)	2	25 × 2	MicroPak® MCH-10 (10)	E-2	0.66	Fl (Ex: 253, Fl: 390)	3	a (4.5)	b	2

Note: E-1 = Acetonitrile + water + acetic acid.
50 : 47 : 3

E-2 = Water + methanol + pentane sodium sulfate.
510 : 490 : (0.01 moles)

[a] 4-(4-Amino-6,7,8-trimethoxy-2-quinazolinyl)-1- piperazine carboxylic acid.
[b] Retention times of other cardiovascular drugs are given.

REFERENCES

1. **Twomey, T. M. and Hobbs, D. C.,** *J. Pharm. Sci.*, 67, 1468, 1978.
2. **Yee, Y. G., Rubin, P. C., and Meffin, P.,** *J. Chromatogr.*, 172, 313, 1979.

PREDNIMUSTINE

High-Presure Liquid Chromatography

Specimen (mℓ)	S	Column cm × mm	Packing (μm)	Elution	Flow rate (mℓ/min)	Det. (nm)	RT (min)	Internal standard (RT)	Other compounds (RT)	Ref.
B (1)	1	30 × 4	μ-Bondapak® C_{18} (10)	E-1 Gradient	2	UV (254, 280)	12	—	Chlorambucil (8) Phenyl acetic mustard (6)	1

Note: E-1 = i. Methanol + 0.175 acetic acid to ii. methanol over 10 min.
69 : 40

REFERENCE

1. Newell, D. R., Hart, L. I., and Harrap, K. R., *J. Chromatogr.*, 164, 114, 1979.

PRIMIDONE

Gas Chromatography

Specimen (mℓ)	S	Column m × mm	Packing (mesh)	Oven temp (°C)	Gas (mℓ/min)	Det.	RT (min)	Internal standard (RT)	Deriv.	Other compounds (RT)	Ref.
B (0.1)	2	1 × 4	3% OV-1 Gas Chrom® Q (100/120)	220	Methane 5 + Argon 95 (60)	ECD	2.5	*p*-Methylprimidone (3.2)	Pentafluorobenzoyl	—	1
B (0.5)	2	2 × 2	3% OV-17 Gas Chrom® Q (80/100)	240	Nitrogen (25)	NPD	4.8	*p*-Methylprimidone (6.5)	—	—	2
B (5)	2	1.5 × 2	3% OV-1 Chromosorb® W (80/100)	Temp. Progr.	Helium (20)	MS-EI	NA	Phenobarbital (NA)	—	2-Ethyl-2-phenylmalonamide (NA)	3
B (0.2) Tissue	2	1.2 × 2	2% SP-2510DA Supelcoport (100/120)	Temp. Progr.	Nitrogen (50)	FID	7.5	*p*-Methylprimidone (9.5) + 2-Ethyl-2-*p*-tolylmalonamide (3.5) + Alphenal (5)	—	Phenobarbital (4.5) 2-Ethyl-2-phenylmalonamide (3)	4
B (0.5)	2	1 × 4	3% OV-1 Gas Chrom® Q (100/120)	135	Methane 5 + Argon 95 (60)	ECD	—	2-Ethyl-2-*p*-tolylmalonamide (2.5)	Reaction with trifluoroacetic anhydride[a]	2-Ethyl-2-phenylmalonamide (1.8)	5

Thin-Layer Chromatography

Specimen (mℓ)	S	Plate (manufacturer)	Layer (mm)	Solvent	Post-separation treatment	Det. (nm)	R_f	Internal standard (R_f)	Other compounds (R_f)	Ref.
B (2)	1	20 × 20 cm (Brinkman)	Silica gel (NA)	S-1	—	UV-reflectance (220)	0.42	—	Phenobarbital (0.67)	6

2-Ethyl-2-phenyl-
malonamide
(0.53)

Note: S-1 = Ethyl acetate + benzene + acetic acid.
90 : 20 : 10

[a] One amide group of malonamides loses a molecule of water to produce nitriles.

REFERENCES

1. Wallace, J. E., Hamilton, H. E., Shimek, Jr. E. L., Schwertner, H. A., and Blum, K., *Anal. Chem.*, 49, 903, 1977.
2. Gupta, R. N., Dobson, K., and Keane, P. M., *J. Chromatogr.*, 132, 140, 1977.
3. Pirl, J. N., Spikes, J. J., and Fitzloff, J., *J. Anal. Toxicol.*, 1, 200, 1977.
4. Leal, K. W., Wilensky, A. J., and Rapport, R. L., *J. Anal. Toxicol.*, 2, 214, 1978.
5. Wallace, J. E., Hamilton, H. E., Shimek, Jr. E. L., Schwertner, H. A., and Haegele, K. D., *Anal. Chem.*, 49, 1969, 1977.
6. Garceau, Y., Philopoulos, Y., and Hasegawa, J., *J. Pharm. Sci.*, 62, 2032, 1973.

PROBENECID

Gas Chromatogaphy

Specimen (mℓ)	S	Column m × mm	Packing (mesh)	Oven temp (°C)	Gas (mℓ/min)	Det.	RT (min)	Internal standard (RT)	Deriv.	Other compounds (RT)	Ref.
B (1)	1	1.8 × NA	1% OV-17 Supelcoport (90/100)	225	Helium (48)	FID	4	*N,N*-Dibutyl analog (6.5)	Methyl[a]	—	1
U (2)	2	1.8 × NA (Steel)	10% OV-1 Chromosorb® W (80/100)	250	Nitrogen (23)	FID	3.9	*N,N*-Dibenzyl-2,5-dimethylbenzenesulfonamide (12.8)	Propyl[b]	R-N(Pr)CH_2CH_2-COOH[c] (11.1) R-N(Pr)CH_2-CH_2-CH_2OH (8) R-N(Pr)CH_2-$CHOHCH_3$ (5.7) R-NH(Pr) (3.1)	2
CSF (0.5)	2	1.8 × 2	3% OV-17 Gas Chrom® Q (80/100)	190	Helium (25)	MS-EI	1.9	*m*-(Di-isobutylsulfamyl) benzoic acid (2)	Pentafluoropropyl	—	3

High-Pressure Liquid Chromatography

Specimen (mℓ)	S	Column cm × mm	Packing (μm)	Elution	Flow rate (mℓ/min)	Det. (nm)	RT (min)	Internal standard (RT)	Other compounds (RT)	Ref.
B (1)	2	25 × 4	Partisil®-ODS (10)	E-1	1.0	UV (252.5)	3.5	—	R-N(Pr)CH_2CH_2-COOH[c] (1.7) R-N(Pr)CH_2CH_2-OH_2CH_3 (2.2) R-N(Pr)CH_2 CH-OH-CH_3 (2.2) R-NH-(Pr) (2.3)	4

Note: E-1 = Acetonitrile + 14 m*M* phosphate buffer, pH 6.
30 : 70

[a] With diazomethane.
[b] With diazopropane.
[c] R = -SO_2-C_6H_4-COOH; Pr = *n*-C_3H_7.

REFERENCES

1. **Zacehei, A. G. and Weidner, L.,** *J. Pharm. Sci.,* 62, 1972, 1973.
2. **Conway, W. D. and Melethil, S.,** *J. Chromatogr.,* 115, 222, 1975.
3. **Faull, K. F., DoAmarall, J. R., and Barchas, J. D.,** *Biomed. Mass Spectrom.,* 5, 317, 1978.
4. **Harle, R. K. and Cowen, T.,** *Analyst,* 103, 492, 1978.

PROCAINAMIDE

Gas Chromatography

Specimen (mℓ)	S	Column m × mm	Packing (mesh)	Oven temp (°C)	Gas (mℓ/min)	Det.	RT (min)	Internal standard (RT)	Deriv.	Other compounds (RT)	Ref.
U (0.1)	2	2 × 1.5	0.75% OV-17 Chromosorb® G AW DMCS (80/100)	270	Nitrogen (25)	FID	2	4-Amino-*N*-(2-piperidion-ethyl) benzamide (4)	—	*N*-Acetylprocainamide (5.5)	1
B, U (1,0.1)	2	0.9 × 2	10% OV-17 Gas Chrom® Q (60/80)	245	Helium (40)	FID	3	4-Amino-*N*-(2-dipropylamino-ethyl) benzamide (4)	—	*N*-Acetylprocainamide (9)	2
B, U (1)	2	1.8 × 2	3% OV-7 Gas Chrom® Q (100/120)	260	Nitrogen (30)	FID	7.5	4-Amino-*N*-(2-dipropylamino-ethyl) benzamide (11)	—	*N*-Acetylprocainamide (23.5)	3
B (1)	2	1.8 × 1	3% OV-17 Chromosorb® W-HP (NA)	240	Nitrogen (40)	FID	6	4-Amino-*N*-(2-dipropylamino-ethyl) benzamide (8)	—	*N*-Acetylprocainamide (20)	4

High-Pressure Liquid Chromatography

Specimen (mℓ)	S	Column cm × mm	Packing (μm)	Elution	Flow rate (mℓ/min)	Det. (nm)	RT (min)	Internal standard (RT)	Other compounds (RT)	Ref.
B (0.5)	2	30 × 4	μ-Bondapak® C_{18} (10)	E-1	2.0	UV (254)	6	*N*-Formyl-procainamide (9)	*N*-Acetyl-procainamide (12)	5
B (0.2)	2	30 × 4	μ-Bondapak® C_{18} (10)	E-2	2.0	UV (280)	3	—	*N*-Acetyl-procainamide (4.5)	6
B (0.5)	2	30 × 4	μ-Bondapak® C_{18} (10)	E-3	2.0	UV (280)	2	*N*-Propionyl-procainamide	*N*-Acetyl-procainamide (2.5)	7

								(3.2)		
B (1)	2	25 × 2.1	Zorbax Sil® (NA)	E-4	1.9	UV (254)	6	*p*-Nitro-(2-diethyl-aminoethyl) benzamide (3)	*N*-Acetyl-procainamide (4.2)	8
B (0.05)	2	30 × 4	μ-Bondapak® CN (10)	E-5	5.0	UV (280)	2.2	—	*N*-Acetyl-procainamide ([illegible].5)	9
B (2)	2	25 × 2.1	LiChrosorb® Si 60 (5)	E-6	1.0	UV (254)	4.5	Pheniramine (3.5)	*N*-Acetyl-procainamide (5.5)	10
B (0.1)	2	NA	Partisil® PXS/SCX (10—25)	E-7	3.0	UV (274)	7.1	*p*-Amino-*N*-(2-dipropylaminoethyl) benzamide (8.7)	*N*-Acetyl-procainamide (5.1)	11
B (0.2)	2	15 × 4.5	Partisil® 5 (5)	E-8	1.0	UV (280)	5.5	—	*N*-Acetyl-procainamide (7.5)	12[a]

Thin-Layer Chromatography

Specimen (mℓ)	S	Plate (manufacturer)	Layer (mm)	Solvent	Post-separation treatment	Det. (nm)	R_f	Internal standard (R_f)	Other compounds (R_f)	Ref.
B (2)	2	NA (Merck)	Silica gel F_{254}	S-1	—	Fl quenching (260)	0.43	*p*-Nitro-*N*-(2-diethylaminoethyl) benzamide (0.55)	*N*-Acetyl-procainamide (0.38)	13
B (0.5—1)	2	20 × 20 (Merck)	Silica gel (0.25)	S-2	—	Reflectance (275)	NA	—	*N*-Acetyl-procainamide (NA)	14
B (0.5)	2	20 × 20 (Merck)	Silica gel (0.25)	S-3	i.— ii. E: HCl	Fl (Ex: i = 313, ii = 297; Fl: 365)	0.32	(i) *p*-Amino-*N*-(2-dipropylaminoethyl)-benzamide (0.74) (ii) *N*-Acetyl *p*-amino-N-(2-dipropylamino ethyl)-benzamide (0.62)	*N*-Acetyl-procainamide (0.23)	15

PROCAINAMIDE (continued)

Note: E-1 = Acetonitrile + acetic acid + water + sodium acetate.
50 mℓ : 40 mℓ : 1 ℓ : 4 g

E-2 = Acetonitrile + 0.75 *M* sodium acetate buffer, pH 3.4.
10 : 90

E-3 = Methanol + water + acetic acid, pH 5.5.
40 : 60 : 1

E-4 = Methanol + water + morpholine.
100 : 2 : 0.1

E-5 = Methanol + 2-propanol + hexane + 10% ammonium hydroxide.
300 : 100 : 100 : 0.1

E-6 = Acetonitrile + 0.1 *M* ammonium acetate + acetic acid.
80 : 20 : 0.1

E-7 = Acetonitrile + 0.1 *M* ammonium phosphate (monobasic) acidified with 0.2% phosphoric acid.
18 : 82

E-8 = Methanol + methylene chloride + methylene chloride saturated with 25% ammonia.
10 : 20 : 70

S-1 = Benzene + 28% ammonia + acetone + dioxane.
10 : 4 : 60 : 20

S-2 = Benzene + 28% ammonia + dioxane.
10 : 15 : 80

[a] Conditions for the analysis of other antiarrhythmic drugs are described.

REFERENCES

1. Karlsson, E., Molin, L., Norlander, B., and Sjöqvist, F., *Br. J. Clin. Pharmacol.*, 1, 467, 1974.
2. Simons, K. J. and Levy, R. H., *J. Pharm. Sci.*, 64, 1967, 1975.
3. Galeazzi, R. L., Sheiner, L. B., Lockwood, T., and Benet, L. Z., *Clin. Pharmacol. Therap.*, 19, 55, 1976.
4. Giardina, E. V., Stein, R. M., and Bigger, J. T., *Circulation*, 55, 388, 1977.
5. Carr, K., Woosley, R. L., and Oates, J. A., *J. Chromatogr.*, 129, 363, 1976.
6. Shukur, L. R., Powers, J. L., Marques, R. A., Winter, M. E., and Sadée, W., *Clin. Chem.*, 23, 636, 1977.
7. Rocco, R. M., Abbot, D. C., Giese, R. W., and Krager, B. L., *Clin. Chem.*, 23, 705, 1977.
8. Dutcher, J. S. and Strong, J. M., *Clin.Chem.*, 23, 1318, 1977.
9. Weddle, O. H. and Mason, W. D., *J. Pharm. Sci.*, 66, 874, 1977.
10. Butterfield, A. G., Copper, J. K., and Midha, K. K., *J. Pharm. Sci.*, 67, 839, 1978.
11. Nation, R. L., Lee, M. G., Huang, S.-M., and Chiou, W. L., *J. Pharm. Sci.*, 68, 532, 1979.
12. Lagerstrom, P.-O., and Persson, B.-A., *J. Chromatogr.*, 149, 331, 1978.
13. Reidenberg, M. M., Drayer, D. E., Levy, M., and Warner, H., *Clin. Pharmacol. Therap.*, 17, 722, 1975.
14. Wesley-Hadzija, B. and Mattocks, A. M., *J. Chromatogr.*, 143, 307, 1977.
15. Gupta, R. N., Eng, F., and Lewis, D., *Anal. Chem.*, 50, 197, 1978.

PROCAINE

High-Pressure Liquid Chromatography

Specimen (mℓ)	S	Column cm × mm	Packing (μm)	Elution	Flow rate (mℓ/min)	Det. (nm)	RT (min)	Internal standard (RT)	Other compounds (RT)	Ref.
D	3	30 × 4	μ-Bondapak® C_{18} (10)	E-1	3	UV (254)	14.3	Pyrrocaine (5.6)	—	1

Note: E-1 = Acetonitrile + water + ammonia.
60 : 40 : 0.01

REFERENCE

1. **Khalil, S. K. W. and Shelver, W. H.,** *J. Pharm. Sci.*, 65, 606, 1976.

PROCARBAZINE

High-Pressure Liquid Chromatography

Specimen (mℓ)	S	Column cm × mm	Packing (μm)	Elution	Flow rate (mℓ/min)	Det. (nm)	RT (min)	Internal standard (RT)	Other compounds (RT)	Ref.
D	3	25 × NA	Partisil PXS 10/25 ODS-2 (10)	E-1	1.5	UV (254)	3.3	Cinnamyl alcohol (16.9)	a	1

Note: E-1 = Methanol + water + ammonium phosphate, pH 5.5.
440 : 560 : 0.05 mole

[a] Retention times of different degradation products of procarbazine are given.

REFERENCE

1. Bruce, G. L. and Boehlert, J. P., *J. Pharm. Sci.*, 67, 424, 1978.

PROCETOFENIC ACID

Gas Chromatography

Specimen (mℓ)	S	Column m × mm	Packing (mesh)	Oven temp (°C)	Gas (mℓ/min)	Det.	RT (min)	Internal standard (RT)	Deriv.	Other compounds (RT)	Ref.
B (0.5)	1	1.5 × 2	4% SE-30 Gas Chrom® Q (80/100)	230	Nitrogen (45)	ECD	4.3	Procetofene (5.3)	Methyl[a]	—	1

[a] With diazomethane.

REFERENCE

1. Desager, J. P., *J. Chromatogr.*, 145, 160, 1978

PROPANTHELINE BROMIDE

Gas Chromatography

Specimen (mℓ)	S	Column m × mm	Packing (mesh)	Oven temp (°C)	Gas (mℓ/min)	Det.	RT (min)	Internal standard (RT)	Deriv.	Other compounds (RT)	Ref.
B, U (910)	2	0.5 × 4	2.6% OV-17 Gas Chrom® Q (80/100)	240	Helium (20)	MS-CI (Ammonia or Methane)	NA	[$^2H_3$]-Propantheline (NA)	—	—	1

REFERENCE

1. Ford, G. C., Grigson, S. J. W., Haskins, N. J., Palmer, R. F., Prout, M., and Vose, C. W., *Biomed. Mass Spectrom.*, 4, 94, 1977.

PROPILDAZINE

Gas Chromatography

Specimen (mℓ)	S	Column m × mm	Packing (mesh)	Oven temp (°C)	Gas (mℓ/min)	Det.	RT (min)	Internal standard (RT)	Deriv.	Other compounds (RT)	Ref.
B (1)	1	2 × 3	3% OV-17 Chromosorb® W (80/100)	Temp. Progr.	Argon 90 + Methane 10 (38)	ECD	9.5	2-(Isopropylamino)-1-(4-nitrophenyl) ethanol (3.6)	Heptafluorobutyryl	—	1

REFERENCE

1. Ventura, P., Zanol, M., Visconti, M., and Pifferi, G., *J. Chromatogr.*, 161, 327, 1978.

PROPIONYLPROMAZINE

Gas Chromatography

Specimen (mℓ)	S	Column m × mm	Packing (mesh)	Oven temp (°C)	Gas (mℓ/min)	Det.	RT (min)	Internal standard (RT)	Deriv.	Other compounds (RT)	Ref.
B, U (5) Tissue	2	2 × 2	3% OV-1 Chromosorb® W (80/100)	250	Nitrogen (35)	a	9	—	—	Promethazine (3.3) Chlorpromazine (5)	1

[a] Flame photometric detector with a 394-nm filter.

REFERENCE

1. Laitem, L., Bello, I., and Gaspar, P., *J. Chromatogr.*, 156, 327, 1978.

PROPOXYPHENE

Gas Chromatography

Specimen (mℓ)	S	Column m × mm	Packing (mesh)	Oven temp (°C)	Gas (mℓ/min)	Det.	RT (min)	Internal standard (RT)	Deriv.	Other compounds (RT)	Ref.
B (5)	1	1.2 × 2 (Steel)	3% OV-225 Gas Chrom® Q (80/100)	185	Nitrogen (50)	ECD	5.8	Imipramine (8)	—	—	1
B (5)	1	1.5 × 2	3% OV-17 Chromosorb® W (80/100)	245	Helium (30)	FID	2.3	Androsterone (7.5)	—	—	2
B (1—4)	1	1.8 × 2	3% SE-30 Gas Chrom® Q (80/100)	216	Helium (33)	FID	3.4	β-Diethylaminoethyldipropyl acetate[a] (5.2)	—	Norpropoxyphene (10.1)[b]	3
B (4)	1	0.6 × 3	2% OV-7 Chromosorb® W (80/100)	Temp. Progr.	Helium (60)	FID	2.9	Pyrroliphene (6.1)	—	Norpropoxyphene (10)[b]	4
B (5—10), U (2)	2	0.5 × 3	1% SE-30 Chromosorb® 750 (100/120)	Temp. Progr.	Nitrogen (40)	FID	8	Cocaine (11)	Nitrosation[c]	Norpropoxyphene (21)	5
B, U (10) Tissue	3	1.8 × 2	3% OV-17 Chromosorb® G (80/100)	250	Nitrogen (16)	FID	5	—	Hydrolysis	Norpropoxyphene (6.5)	6
B (5) Tissue	3	0.9×NA	3.8% SE-30 Chromosorb® W (80/100)	225	Helium (d)	FID	1.2	Pyrroliphene (2.2)	—	—	7
B, U (20) Tissue	3	1.8 × 3	3% SE-30 Chromosorb® W (80/100)	210	Nitrogen (60)	FID	2.3	—	Trimethylsilyl	Norpropoxyphene (6.5)[b]	8
B (5)	2	1.8 × 2	3% OV-27 Chromosorb® W (80/100)	180	Nitrogen (40)	FID	8	Promethazine (13)	$LiAlH_4$-Reduction	Norpropoxyphene (20.3)	9
B (2.5)	2	1.8 × 2	2.8% OV-210 +	Temp.	Helium (40)	NPD	1.8	Pyrroliphene (3)	Reduction[e]	Norpropoxyphene	10

		3.2% OV-1 Chromosorb® W (80/100)	Progr.		(2.4)	

High-Pressure Liquid Chromatography

Specimen (mℓ)	S	Column cm × mm	Packing (μm)	Elution	Flow rate (mℓ/min)	Det. (nm)	RT (min)	Internal standard (RT)	Other compounds (RT)	Ref.
D	3	25 × 2	LiChrosorb® (10)	E-1	1.5	UV (254)	2.5	—	—	11

Thin-Layer Chromatography

Specimen (mℓ)	S	Plate (manufacturer)	Layer (mm)	Solvent	Post-separation treatment	Det. (nm)	R_f	Internal standard (R_f)	Other compounds (R_f)	Ref.
B, U Tissue	3	5 × 20 cm (NA)	Silica gel (0.25)	S-1	Sp: Acidified iodoplatinate reagent	Visual	0.89 0.84[f]	—	Norpropoxyphene (0.60)	12

Note: E-1 = Hexane + isopropanol containing 5% ammonia.
99.35 : 0.65

S-1 = Methanol + ammonia.
100 : 1.5

[a] SKF 525-A obtained from Smith, Kline, and French laboratories.
[b] Norpropoxyphene is converted to its amide during extraction.
[c] Only norpropoxyphene undergoes nitrosation producing *N*-nitrosonorpropoxyphene.
[d] Inlet pressure = 50 psig.
[e] With sodium bis-(2-methoxyethoxy) aluminum dihydride.
[f] R_f of hydrolyzed propoxyphene.

PROPOXYPHENE (continued)

REFERENCES

1. Maynard, W. R., Jr., Bruce, R. B., and Fox, G. G., *Anal. Lett.*, 6, 1005, 1973.
2. Evenson, M. A. and Koellner, S., *Clin. Chem.*, 19, 492, 1973.
3. Verebely, K. and Inturrisi, C. E., *J. Chromatogr.*, 75, 195, 1973.
4. Nash, J. F., Bennet, I. F., Bopp, R. J., Brunson, M. K., and Sullivan, H. R., *J. Pharm. Sci.*, 64, 429, 1975.
5. Serfontein, W. J. and de Villiers, L. S., *J. Pharm. Pharmacol.*, 28, 718, 1976.
6. Norheim, G., *Arch. Toxicol.*, 36, 89, 1976.
7. Rejent, T. A., Michalek, R. W., and Lehotay, J. M., *Clin. Toxicol.*, 11, 43, 1977.
8. Christensen, H., *Acta Pharmacol. Toxicol.*, 40, 289, 1977.
9. Cleemann, M., *J. Chromatogr.*, 132, 287, 1977.
10. Angelo, H. R. and Christenson, J. M., *J. Chromatogr.*, 140, 280, 1977.
11. Gilpin, R. K., Korpi, J. A., and Janicki, C. A., *J. Chromatogr.*, 107, 115, 1975.
12. McBay, A. J., Turk, R. F., Corbett, B. W., and Hudson, P., *J. Forensic Sci.*, 19, 81, 1974.

PROPRANOLOL

Gas Chromatography

Specimen (mℓ)	S	Column m × mm	Packing (mesh)	Oven temp (°C)	Gas (mℓ/min)	Det.	RT (min)	Internal standard (RT)	Deriv.	Other compounds (RT)	Ref.
B (0.5—2)	1	2 × 4	3% OV-17 Chromosorb® Q (100/120)	205	Nitrogen (70)	ECD	3.8	Pronethalol (1.9)	Heptafluorobutyryl	—	1
Brain tissue	1	1.8 × 2	1% OV-17 + 2% OV-1 Chromosorb® W (80/100)	170	Nitrogen (30)	ECD	9.8	Oxprenolol (3.9)	Trifluoroacetyl	Propanololglycol (2.5) N-Desisopropylpropranolol (6.3)	2
B (1)	1	1.8 × 2	10% OV-1 Gas Chrom® Q (80/100)	245	Methane 5 + Argon 95 (20)	ECD	3.2	4-Methylpropranolol (4.2)	Pentafluoropropionyl	—	3
U (20)	2	1.8 × 2	3% SP-2100 Supelcoport (100/120)	Temp. Progr.	Helium (40)	MS-EI	NA	Metatolyloxy analog (NA) + Phenoxyacetic acid (NA)	i. Methyl[a] ii. Trifluoroacetyl[b]	c	4
B (1)	2	1.8 × 2	5% OV-17 Chromosorb® W (80/100)	185	Nitrogen (30)	ECD	—	Ethylnaphthoxylactate (11)	i. Methyl[a] ii. Heptafluorobutyryl	Naphthoxylacetic acid (9)	5
B (0.5—2)	1	2 × 4	3% OV-225 Chromosorb® W (80/100)	250	Nitrogen (60)	ECD	l = 12.5[d] d = 13.5	Diazepam (7.5)	N-Heptafluorobutyryl-1-prolyl	—	6

High-Pressure Liquid Chromatography

Specimen (mℓ)	S	Column cm × mm	Packing (μm)	Elution	Flow rate (mℓ/min)	Det. (nm)	RT (min)	Internal standard (RT)	Other compounds (RT)	Ref.
B (1)	1	30 × 4	μ-Bondapak-CN (10)	E-1	2.0	Fl (Ex: 220 Fl: —)[e]	4.1	—	4-Hydroxypropranolol (6.1)	7

PROPRANOLOL (continued)

High-Pressure Liquid Chromatography

Specimen (mℓ)	S	Column cm × mm	Packing (μm)	Elution	Flow rate (mℓ/min)	Det. (nm)	RT (min)	Internal standard (RT)	Other compounds (RT)	Ref.
B (4)	1	30 × 3.9	μ-Bondapak®-C_{18} (10)	E-2	2.0	Fl (Ex: f, Fl: g)	5.2	N-Ethylpropranolol (7.5)	—	8
B (1)	1	30 × 4	μ-Bondapak® alkylphenyl (10)	E-3	2.0	Fl (Ex: 205, Fl: h)	7.9	4-Methylpropranolol (11.6)	4-Hydroxypropranolol (3.9)	9
U (1)	2	30 × 4	μ-Bondapak® C_{18} (10)	E-4	2.0	UV (295) Fl: i	7.48	Procainamide (3.2)[j] + 2-Maphthoxyacetic acid (5.5)[k]	Propranololglycol (3.4)[j] 4-Hydroxypropranolol (5)[j] 1-Naphthoxylactic acid (4.5)[k] 1-Naphthoxyacetic acid (6.2)[k]	10
B (2)	1	30 × 3.9	μ-Bondapak® C_{18} (10)	E-5	1.0	Fl (Ex: 295, Fl: 360)[l]	6	Pronethalol (8.5) + Labetalol (7)	4-Hydroxypropranolol (5)	11
B (0.5)	1	30 × 4	μ-Bondapak® CN (10)	E-6	2.0	Fl (Ex: 276, Fl: m)	10	Cyclomethycaine (13.5)	4-Hydroxypropranolol (7.7)	12
B (2)	1	25 × 4.6	Partisil®-ODS[n] (10)	E-7	1.5	Fl (Ex: 285, Fl: 350)	6.3	Proethalol (4.8)	—	13

Thin-Layer Chromatography

Specimen (mℓ)	S	Plate (manufacturer)	Layer (mm)	Solvent	Post-separation treatment	Det. (nm)	R_f	Internal standard (R_f)	Other compounds (R_f)	Ref.
B (1), U (4)	2	20 × 20 cm (Merck)	Silica gel 60 (0.25)	S-1	Sp: 10% citric acid in glycol-water (1:1)	Fl (reflectance) (Ex: 290, Fl: 365)	0.42	—	Desisopropylpropranolol (0.17)	14

B, U (0.5—5)	2	10 × 20 cm (Merck)	Silica gel (0.25)	S-2	Sp: Propylene-glycol + water (1:1)	Fl (reflectance) (Ex: 300, Fl: 300—420)	0.34	—	Metabolites[e]	15
B (1)	2	20 × 20 cm (Merck)	Silica gel 60 (0.25)	S-3	—	UV-reflectance (288)	0.53	—	—	16
D	3	NA (Gelman)	Silica gel (NA)	S-4	E: I_2 vapors	Visual; sp	0.51	—	α-Naphthoxylactic acid (0.02) 4-Hydroxypropranolol (0.38) Propranolol glycol (0.66)	17

Note: E-1 = Methanol + 2-propanol + hexane + 2.9 *M* ammonia.
25 : 25 : 50 : 0.04

E-2 = Methanol + water + acetic acid.
70 : 60 : 0.6

E-3 = Acetonitrile + 0.06% phosphoric acid.
27 : 73

E-4 = Acetonitrile + methanol + acetic acid + water (for basic extracts).
35 : 5 : 1 : 59
Acetonitrile + acetic acid + water (for acid extracts).
36 : 1 : 63

E-5 = Methanol + water + acetic acid + (0.005 *M* heptanesulfonic acid).
60 : 39 : 1 :

E-6 = Acetonitrile + 0.02 *M* acetate buffer, pH 7.
70 : 30

E-7 = Acetonitrile + 0.1 *M* phosphate buffer, pH 2.7.
25 : 75

S-1 = Ethyl acetate + benzene + methanol (in ammonia atmosphere).
20 : 20 : 10

PROPRANOLOL (continued)

S-2 = Methanol + ammonia.
100 : 0.4

S-3 = Methanol + ammonia.
100 : 1.5

S-4 = i. Acetonitrile + benzene + hexane + ammonia (development to 10 cm).
80 : 40 : 40 : 1
ii. Hexane + diethyl ether (development to 16 cm).
9 : 1

[a] With diazomethane. Fluorinated derivative was prepared from the dried residue.
[b] A separate portion of the urine extract was subjected to trifluoroacetylation only.
[c] Computer constructed chromatogram of a number of metabolites is given.
[d] Enantiomers of propranolol.
[e] Different wavelength settings are required for the assay of 4-hydroxypropranolol.
[f] Interference filter: 295 nm.
[g] Kodak filter #34.
[h] Filter: KV 340.
[i] High intensity gramicidin UV lamp, primary UV interference filter (310 nm, Baird Atomic).
[j] Analysis of basic extract.
[k] Analysis of acid extract.
[l] Aminco 295 nm primary filter and Corning #7-51 secondary filter. For the analysis of 4-hydroxypropranolol, a Corning #7-51 primary filter and Wratten #2A secondary filter were used.
[m] Cutoff filter type 340.
[n] Column temperature: 50°C.
[o] Conditions for the analysis of basic and acid metabolites are described.
[p] For radioactive compounds, the sheets are cut, and desired spots are eluted to measure radioactivity.

REFERENCES

1. Salle, E. D., Baker, K. M., Bareggi, S. R., Watkins, W. D., Chedsey, C. A., Fugeno, A., and Morselli, P. L., *J. Chromatogr.*, 84, 347, 1973.
2. Saelens, D. A., Walle, T., and Privitera, P. J., *J. Chromatogr.*, 123, 185, 1976.
3. Kates, R. E. and Jones, C. L., *J. Pharm. Sci.*, 66, 1490, 1977.
4. Vu, V. T. and Abramson, F. P., *Biomed. Mass Spectrom.*, 5, 686, 1978.
5. Easterling, D. E., Walle, T., Conradi, E. C., and Gaffney, T. E., *J. Chromatogr.*, 162, 439, 1979.
6. Caccia, S., Guiso, G., Ballabio, M., and de Ponte, P., *J. Chromatogr.*, 172, 457, 1979.
7. Mason, W. D., Amick, E. N., and Weddle, O. H., *Anal. Lett.*, 10, 515, 1977.
8. Wood, A. J. J., Carr, K., Vestal, R. E., Belcher, S., Wilkinson, G. R., and Shanel, D. G., *Br. J. Clin. Pharmacol.*, 6, 345, 1978.
9. Nation, R. L., Peng, G. W., and Chiou, W. L., *J. Chromatogr.*, 145, 429, 1978.
10. Pritchard, J. F., Schneck, D. W., and Hayes, A. H., Jr., *J. Chromatogr.*, 162, 47, 1979; *Res. Commun. Chem. Pathol.*, 23, 279, 1979; 24, 3, 1979.
11. Taburet, A.-M., Taylor, A. A., Mitchell, J. R., Rollins, D. E., and Pool, J. L., *Life Sci.*, 24, 209, 1979.
12. Nygard, G., Shelver, W. H., and Khalil, S. K. W., *J. Pharm. Sci.*, 68, 379, 1979.
13. Jatlow, P., Bush, W., and Hochster, H., *Clin. Chem.*, 25, 777, 1979.
14. Schäffer, M., Geissler, H. E., and Mutschler, E., *J. Chromatogr.*, 143, 607, 1977.
15. Garceau, Y., Davis, I., and Hasegawa, J, *J. Pharm. Sci.*, 67, 826, 1978.
16. Hadzija, B. W. and Mattocks, A. M., *J. Pharm. Sci.*, 67, 1307, 1978.
17. Abou-Donia, M. B., Bakry, N. M., and Strauss, H. C., *J. Chromatogr.*, 172, 463, 1979.

PROPYLTHIOURACIL

High-Pressure Liquid Chromatography

Specimen (mℓ)	S	Column cm × mm	Packing (μm)	Elution	Flow rate (mℓ/min)	Det. (nm)	RT (min)	Internal standard (RT)	Other compounds (RT)	Ref.
B (1)	2	i. 30 × 3.9 ii. 25 × 4.6	i. μ-Bondapak®[a] C_{18} (10) ii. LiChrosorb® RP-18 (10)	E-1	2.1	UV (275)	4.5	—	—	1
B (3)	2	25 × 3.1	Spherisorb RP-18 (10)	E-2	2.3	UV (280)	4.8	Methyluracil (1.5)	—	2

Note: E-1 = Acetonitrile + 0.01 *M* NaH_2PO_4, pH 3.
45 : 100

E-2 = Acetic acid + methanol + water.
10 : 75 : 915

[a] Two columns are combined in series.

REFERENCES

1. Giles, H. G., Miller, R., and Sellers, E. M., *J. Pharm. Sci.,* 68, 1459, 1979.
2. Ringhand, H. P. and Ritschel, W. A., *J. Pharm. Sci.,* 68, 1461, 1979.

PROPYPHENAZONE

Gas Chromatography

Specimen (mℓ)	S	Column m × mm	Packing (mesh)	Oven temp (°C)	Gas (mℓ/min)	Det.	RT (min)	Internal standard (RT)	Deriv.	Other compounds (RT)	Ref.
B (0.2)	1	25 × 0.45[a]	Carbowax® 20 *M*	220	Helium (2.5)	NPD	3.0	Lidocaine (2.3)	—	Caffeine (3.5) Desmethylpropyphenazone (4.2)	1
B (0.5)	1	1 × 3	3% Poly-I 110 Chromosorb® W (80/100)	220	Nitrogen (40)	FID	4	Hexacosane (13)	—	—	2

Thin-Layer Chromatography

Specimen (mℓ)	S	Plate (manufacturer)	Layer (mm)	Solvent	Post-separation treatment	Det. (nm)	R_f	Internal standard (R_f)	Other compounds (R_f)	Ref.
B (0.5)	2	NA (Laboratory)	Silica gel-H F^{254} (NA)	S-1	—	UV (274)[b]	NA	—	—	3

Note: S-1 = Chloroform + methanol + diethylamine.
98 : 1 : 1

[a] Support coated open tubular column.
[b] Spots are visualized under a UV lamp, scraped and eluted with methanol. The absorbance of eluate is determined spectroscopically.

REFERENCES

1. **van den Bosch, N. and de Vos, D.,** *Fresenius Z. Anal. Chem.,* 296, 46, 1979.
2. **Sioufi, A. and Marfil, F.,** *J. Chromatogr.,* 146, 508, 1978.
3. **Dell, H.-D. and Kolle, J.,** *Fresenius Z. Anal. Chem.,* 289, 288, 1978.

PROSTAGLANDINS

Gas Chromatography

Specimen (mℓ)	S	Column m × mm	Packing (mesh)	Oven temp (°C)	Gas (mℓ/min)	Det.	RT (min)	Internal standard (RT)	Deriv.	Other compounds (RT)	Ref.
Semen (5—10)	2	2 × 2.5	3% OV-17 Gas Chrom® Q (100/120)	230	Helium (45)	FID	E = 11 $F_{2\alpha}$ = 11 $F_{1\alpha}$ = 12 B = 20.5 A = 36.5	—	Trimethyl-silyl	—	1
Semen (130) Tissue	2	3 × 4	1% OV-1 Gas Chrom® Q (100/120)	245	Nitrogen (50)	FID	—	—	i. Methyl ester ii. Tri-tert-butyl-di-methylsilyl ether	Thromboxane B_2 (20)[a]	2
D	2	25[b] × 0.6	HI-EFF-1BP	200	Nitrogen (2.5)	ECD	B_1 = 8.5 B_2 = 9.5	—	i. Methyles-ter ii. Trimeth-ylsilyl ether	—	3
D	2	3 × 2.5	1% OV-101 Gas Chrom® Q (80/100)	240	Helium (40)	FID	D_2 = 31 E_1 = 32.5 E_2 = 30, 31.5 $F_{1\alpha}$ = 35 $F_{2\alpha}$ = 33.5	—	i. Methyl ester ii. Methox-imes iii. Dimeth-ethylsilyl ether	—	4
D	2	1.5 × NA	1% OV-1 Supelcoport (60/80)	235	Helium (20)	MS-EI	E_1 = 3.5	$[^2H_4]$-Prosta-glandins (E_1 = 3.5)	i. Methyl ester ii. Methox-imes iii. *O*-Ace-tyl	—	5

D	2	1 × 3	3% OV-101 Celite® 545 (80/100)	250	Helium (30)	MS-CI	NA[c]	—	i. Methioxime ii. Trimethylsilyl	—	6
D	2	1.5 × 2	1% OV-225 Gas Chrom® Q (100/120)	195	Helium (20)	MS-EI	$F_{1\alpha}$ = 6 $F_{2\alpha}$ = 5.5	[$^2H_4$] $F_{2\alpha}$ Prostaglandin (5.5)	i. Methyl ester ii. Trimethylsilyl ether	—	7
CSF	2	1.5 × NA	1% SE-30 NA	235	Helium (20)	MS-EI	NA[d]	[^2H]-Prostaglandins (NA)	i. Mehoximes ii. Methyl ester iii. Trimethylsilyl ether	—	8
U (NA)	2	30[b] × 0.3	SE-30	NA	Helium (2)	MS-EI	NA	[^2H]-Prostaglandins (NA)	i. Methyl ester ii. Methoximes iii. Trimethylsilyl ether	—	9
Cell preparation	2	25[b] × NA	Methyl phenyl polysiloxane	250	Helium (e)	ECD	F_2 = 30 $F_{2\alpha}$ = 26 6-Keto-$F_{1\alpha}$ = 36	—	i. Pentafluoroethyl ii. Methoxime iii. Trimethylsilyl ether	Thromboxane B_2 (32)	10

High-Pressure Liquid Chromatography

Specimen (mℓ)	S	Column cm × mm	Packing (μm)	Elution	Flow rate (mℓ/min)	Det. (nm)	RT (min)	Internal standard (RT)	Other compounds (RT)	Ref.
D	3	1 × 2.1	Vydac®[f] cation	E-1	0.5	UV (254)[g]	E_2 = 10	—	—	11

PROSTAGLANDINS (continued)

High-Pressure Liquid Chromatography

Specimen (mℓ)	S	Column cm × mm	Packing (μm)	Elution	Flow rate (mℓ/min)	Det. (nm)	RT (min)	Internal standard (RT)	Other compounds (RT)	Ref.
			exchange resin (30—44)				Trans E_2 = 7			
D	3	1 × 2.1	LiChrosorb® SI60 (20)	E-2	1	UV (254)[g]	F_2 = NA 15 epi F_2 = NA	Diphenylurea ester of cholic acid (NA)	—	12
D	3	50 × 2.3	Vydac® anion exchange resin (NA)	E-3	0.5	UV (Variable)[h]	E = 4 A = 6 B = 10	—	—	13
D	3	100 × NA	Silica gel-ODS (37)	E-4	i	UV (254)[j]	Cis A_2 = 8.9 Trans A_2 = 17 5,6-Cis B_2 = 16.4 5,6-Trans B_2 = 15.8	—	—	14
D	2	60 × 4	μ-Bondapak® C_{18} (10)	E-5	0.75	UV (254)[k]	E_2 = 7.4 E_1 = 8.5 A_2 = 12.3 A_1 = 14.4 B_2 = 25.9 B_1 = 30	—	—	15
Tissue	2	NA	MicroPak CN (10)	E-6 Gradient	1.7	Fl[m] (Ex: 325, Fl: 385)	D_2 = 11 E_2 = 14.5 $F_{2\alpha}$ = 23 6-Keto-$F_{1\alpha}$ = 30 6-Hydroxy-$F_{1\alpha}$ = 39	—	Thromboxane B_2 (19)	16
U (NA) tissue	2	30 × 4	μ-Bondapak® Fatty acid (10)	E-7	2	n	6-Keto-$F_{1\alpha}$ = 10 $F_{2\alpha}$ = 21 D_2 = 30 E_2 = 30	—	Thromboxane B_2 (18)	17

D	2	25 × 4	Partisil®-ODS[o] (10)	E-8	3.6	UV (205)	E_1 = 41 I_2 = 6	—	—	18
D	2	30 × 4	μ-Bondapak® C_{18} (10)	E-9	1.0	UV (205)	I_2 = 7	—	—	19

Thin-Layer Chromatography

Specimen (mℓ)	S	Plate (manufacturer)	Layer (mm)	Solvent	Post-separation treatment	Det. (nm)	R_f	Internal standard (R_f)	Other compounds (R_f)	Ref.
Tissue	2	NA (Brinkman)	Silica gel F_{254} (NA)	S-1	—	Fl[m] (348)	$D_2 = 0.56$ $E_2 = 0.42$ $F_{2\alpha} = 0.23$ 6-Keto-$F_{1\alpha} = 0.18$ 6-Hydroxy-$F_{1\alpha} = 0.10$	—	Thromboxane B_2 (0.32)	16
Semen	2	NA (Merck)	Silica gel F_{254} (NA)	S-2[p]	Sulfuric acid charring	Visual[q]	$E_1 = 0.22$ 8-Iso-$E_1 = 0.30$ 8-Iso-$E_2 = 0.31$ $F_{2\alpha} = 0.10$ 8-Iso-$F_{1\alpha} = 0.10$ 8-Iso-$F_{2\alpha} = 0.09$ 19-Hydroxy-$E_2 = 0.04$	—	—	20

Note: E-1 = A + B (55 : 45); A = Chloroform + hexane (2 : 1); B = A (17) + acetonitrile (17 : 3).

E-2 = Acetonitrile + methylene chloride + water (50 : 50 : 1).

PROSTAGLANDINS (continued)

E-3 = 0.1 *M* Phosphate buffer, pH 6 + 0.5% β-cyclodextrin.

E-4 = Methanol + water + silver prechlorate.
200 : 800 : 0.5 moles

E-5 = Acetonitrile + water.
85 : 15

E-6 = i. Chloroform + isooctane + methanol.
35 : 65 : 1
ii. Chloroform + methanol.
100 : 13.5

E-7 = Water + acetonitrile + benzene + acetic acid.
76.7 : 23 : 0.2 : 0.1

E-8 = Water + methanol + boric acid + sodium tetraborate.
666 : 333 : 2.5 g : 3.8 g

E-9 = Acetonitrile + water, pH 9.3 (with borate buffer).
20 : 80

S-1 = Chloroform + methanol.
100 : 7

S-2 = Ethyl acetate + acetone + acetic acid.
90 : 14 : 1

[a] A number of alternative derivatives have been prepared and analyzed by GC and GC-MS.
[b] Wall coated open tubular capillary column.
[c] Each prostaglandin of A and E series produced two peaks.
[d] The extract of CSF is separated by thin-layer chromatography on silica gel plates using ethyl acetate – acetic acid – trimethyl pentane – water (110:20:30:100) as the developing solvent. Each spot is scraped, eluted, derivatized, and quantitated by GC-MS.
[e] 22.5 cm/sec.
[f] Treated with 0.6 *M* solution of silver nitrate. The column is maintained at 26°C.
[g] *p*-Nitrophenacyl esters are prepared prior to chromatography.
[h] E = 210 nm; A = 221 nm; B = 282 nm.
[i] 900 psi.

[j] Methyl esters were chromatographed.
[k] p-Nitrobenzyloximes were prepared of methyl esters of prostaglandins prior to chromatography.
[l] Retention data for p-nitrobenzyloximes of other prostaglandins—methyl esters; of prostaglandins—pentafluorobenzyl esters, and pentafluorobenzyloximes of prostaglandins—pentafluorobenzyl esters, are also given.
[m] Prostaglandins were derivatized with 4-bromomethyl-7-methoxycoumarin to form fluorescent derivative.
[n] Fractions were collected to measure radioactivity; separated fractions were analyzed by GC-MS.
[o] Prepared and specially deactivated by the authors.
[p] Group separation of biological extracts is carried by chromatography on Sephadex® LH-20 columns.
[q] Separated spots are eluted, derivatized, and analyzed by GC-MS.

REFERENCES

1. **Rosello, J., Tusell, J., and Gelpi, E.,** *J. Chromatogr.*, 130, 65, 1977.
2. **Smith, A. G., Harland, W. A., and Brooks, C. J. W.,** *J. Chromatogr.*, 142, 533, 1977.
3. **Korteweg, M., Verdonk, G., Sandra, P., and Verzele, M.,** *Prostaglandins*, 13, 1221, 1977.
4. **Miyazaki, H., Ishibashi, M., Yamashita, K., and Katori, M.,** *J. Chromatogr.*, 153, 83, 1978.
5. **Goldyne, M. E. and Hammarström, S.,** *Anal. Biochem.*, 88, 675, 1978.
6. **Ariga, T., Suzuki, M., Morita, I., Musora, S.-I., and Miyatake, T.,** *Anal. Biochem.*, 90, 174, 1978.
7. **Ferretti, A. and Flanagan, V. P.,** *Anal. Lett.*, B11, 195, 1978.
8. **Abdel-Halim, M. S., Ekstedt, J., and Änggard, E.,** *Prostaglandins*, 17, 405, 1979.
9. **Erlenmaier, T., Müller, H., and Seyberth, H. W.,** *J. Chromatogr.*, 163, 289, 1979
10. **Fitzpatrick, F. A., Stringfellow, D. A., MacLouf, J., and Rigaud, M.,** *J. Chromatogr.*, 177, 51, 1979.
11. **Merritt, M. V. and Bronson, G. E.,** *Anal. Chem.*, 48, 1851, 1976.
12. **Roseman, T. J., Butler, S. S., and Douglas, S. L.,** *J. Pharm. Sci.*, 65, 673, 1976.
13. **Uekama, K., Hirayama, F., Ikeda, K., and Inaba, K.,** *J. Pharm. Sci.*, 66, 706, 1977.
14. **Weber, D. J.,** *J. Pharm. Sci.*, 66, 744, 1977.
15. **Fitzpatrick, F. A., Wynalda, M. A., and Kaiser, D. G.,** *Anal. Chem.*, 49, 1032, 1977.
16. **Turk, J., Weiss, S. J., Davis, J. E., and Needleman, P.,** *Prostaglandins*, 16, 291, 1978.
17. **Whorton, A. R., Carr, K., Smigel, M., Walker, L., Ellis, K., and Oates, J. A.,** *J. Chromatogr.*, 163, 64, 1979; *Anal. Biochem.*, 98, 455, 1979.
18. **Hill, G. T.,** *J. Chromatogr.*, 176, 407, 1979.
19. **Wynalda, M. A., Lincoln, F. H., and Fitzpatrick, F. A.,** *J. Chromatogr.*, 176, 413, 1979.
20. **Taylor, P. L.,** *Prostaglandins*, 17, 259, 1979.

PROTRIPTYLINE

Gas Chromatography

Specimen (mℓ)	S	Column m × mm	Packing (mesh)	Oven temp (°C)	Gas (mℓ/min)	Det.	RT (min)	Internal standard (RT)	Deriv.	Other compounds (RT)	Ref.
D	2	1 × 2 (Nickel)	3% OV-101 Chromosorb® W (NA)	200	Helium (30)	MS-EI MS-CI	4.5	—	*N*-Ethyl[a]	Desipramine (9.8) Nortriptyline (8.1) Desmethyldoxepin (4)	1[b]

Thin-Layer Chromatography

Specimen (mℓ)	S	Plate (manufacturer)	Layer (mm)	Solvent	Post-separation treatment	Det. (nm)	R_f	Internal standard (R_f)	Other compounds (R_f)	Ref.
B (1)	2	20 × 20 cm (Merck)	Silica gel (0.25)	S-1	—	Fl (Ex: 297, Fl: 365)	0.23	—	—	2

Note: S-1 = Methanol + ammonia.
100 : 1.5

[a] With diethylacetal in dimethylformamide.
[b] This paper primarily describes the preparation of *N*-alkyl derivatives of secondary amines.

REFERENCES

1. Narsimhachari, N. and Friedel, R. O., *Anal. Lett.*, 12 B, 77, 1979.
2. Gupta, R. N. and Eng, F., unpublished data.

PROXIPHYLLINE

High-Pressure Liquid Chromatography

Specimen (mℓ)	S	Column cm × mm	Packing (μm)	Elution	Flow rate (mℓ/min)	Det. (nm)	RT (min)	Internal standard (RT)	Other compounds (RT)	Ref.
B (0.5)	2	200 × 2	Zipax® SCX[a] (NA)	E-1	0.57	UV (273)	11	—	b	1
B (0.2)	2	30 × 3.9	μ-Bondapak® C_{18} (10)	E-2	1.6	UV (254, 280)	5.3	8-Chlorotheophylline (7.3)	Theophylline (4) b	2

Note: E-1 = 1 m*M* H_3PO_4, pH 2.9.

E-2 = Methanol + 0.02 *M* potassium chloride, pH 2 (with HCl).
30 : 70

[a] Column temperature = 50°C.
[b] Theophylline and a number of other xanthine metabolites do not interfere.

REFERENCES

1. **Selvig, K. and Bjerve, K. S.,** *Scand. J. Clin. Lab. Invest.*, 37, 373, 1977.
2. **Nielsen-Kudsk, F. and Pedersen, A. K.,** *Acta Pharmacol. Toxicol.*, 42, 298, 1978.

PYRAZINAMIDE

Gas Chromatography

Specimen (mℓ)	S	Column m × mm	Packing (mesh)	Oven temp (°C)	Gas (mℓ/min)	Det.	RT (min)	Internal standard (RT)	Deriv.	Other compounds (RT)	Ref.
B (0.2)	1	1 × 2	3% OV-17 Chromosorb® W (80/100)	Temp. Progr.	Isobutane[a]	MS-CI	2	Nicotinic acid (1.2) Nicotinamide (2.7)	Trimethylsilyl	Pyrazinoic acid (1.4) 5-Hydroxypyrazinoic acid (3.2)	1

[a] Pressure = 1 Torr.

REFERENCE

1. Roboz, J., Suzuki, R., and Yü, T.-F., *J. Chromatogr.*, 147, 337, 1978.

PYRAZOLOIMIDAZOLE

Gas Chromatography

Specimen (mℓ)	S	Column m × mm	Packing (mesh)	Oven temp (°C)	Gas (mℓ/min)	Det.	RT (min)	Internal standard (RT)	Deriv.	Other compounds (RT)	Ref.
B (1)	1	1.8 × 2	3% SP-2100DB Supelcoport (100/120)	175	Nitrogen (30)	NPD	1.8	2,2-Dipyridyl (2.2)	—	—	1

REFERENCE

1. **Ames, M. M., Powis, G., and Kuehn, P.,** *J. Chromatogr.*, 169, 412, 1979.

PYRIDINOL CARBAMATE

High-Pressure Liquid Chromatography

Specimen (mℓ)	S	Column cm × mm	Packing (μm)	Elution	Flow rate (mℓ/min)	Det. (nm)	RT (min)	Internal standard (RT)	Other compounds (RT)	Ref.
B (1)	2	15 × 4.5	LiChrosorb® Si60 (5)	E-1	1.75	UV (254)	3.9	Chlorpromazine (5.3)	—	1

Note: E-1 = Dichloromethane + isopropanol + ammonia.
90 : 10 : 0.1

REFERENCE

1. Bernard, N., Brazier, J. L., and Sassard, J., *J. Chromatogr.*, 152, 260, 1978.

PYRIDOXINE

Gas Chromatography

Specimen (mℓ)	S	Column m × mm	Packing (mesh)	Oven temp (°C)	Gas (mℓ/min)	Det.	RT (min)	Internal standard (RT)	Deriv.	Other compounds (RT)	Ref.
D	3	1.8 × 2	5% DC-550 Chromosorb® P (80/100)	150	Argon (40)	FID	6	—	Trifluoroacetyl[a]	Desoxypyridoxine (5) Pyridoxal (12) Pyridoxamine (14)	1

High-Pressure Liquid Chromatography

Specimen (mℓ)	S	Column cm × mm	Packing (μm)	Elution	Flow rate (mℓ/min)	Det. (nm)	RT (min)	Internal standard (RT)	Other compounds (RT)	Ref.
D	3	30 × 4	μ-Bondapak® C_{18} (10)	E-1	0.5	UV (270)	14	—	Nicotinic acid (7) Nicotinamide (9) Thiamine (27) Riboflavin (32)	2

Note: E-1 = Methanol + 0.005 *M* sodium hexanesulfonate containing 1% acetic acid.
250 : 750

[a] With *N*-methyl-bis-trifluoroacetamide.

REFERENCES

1. **Patzer, E. M. and Hilker, D. M.,** *J. Chromatogr.*, 135, 489, 1977.
2. **Kirchmeier, R. L. and Upton, R. P.,** *J. Pharm. Sci.*, 67, 1444, 1978.

PYRIMETHAMINE

High-Pressure Liquid Chromatography

Specimen (mℓ)	S	Column cm × mm	Packing (μm)	Elution	Flow rate (mℓ/min)	Det. (nm)	RT (min)	Internal standard (RT)	Other compounds (RT)	Ref.
B (2)	1	100 × 4.6	Spherisorb® SSW (5)	E-1	2	UV (254)	3.5	Metoprine (5)	Dapsone (2.5) Monoacetyldapsone (6.2)	1

Thin-Layer Chromatography

Specimen (mℓ)	S	Plate (manufacturer)	Layer (mm)	Solvent	Post-separation treatment	Det. (nm)	R_f	Internal standard (R_f)	Other compounds (R_f)	Ref.
B (1.5) tissue	1	20 × 20 cm (Merck)	Silica gel F_{254} (0.25)	S-1	—	UV reflectance (275)	0.49	—	—	2

Note: E-1 = Di-isopropyl ether + methanol + 21% ammonia.
96 : 4 : 0.1

S-1 = Chloroform + methanol.
70 : 30

REFERENCES

1. Jones, C. R. and Ovenell, S. M., *J. Chromatogr.*, 163, 179, 1979.
2. DeAngelis, R. L., Simmons, W. S., and Nichol, C. A., *J. Chromatogr.*, 106, 41, 1975.

PYRITHIOXIN

High-Pressure Liquid Chromatography

Specimen (mℓ)	S	Column cm × mm	Packing (μm)	Elution	Flow rate (mℓ/min)	Det. (nm)	RT (min)	Internal standard (RT)	Other compounds (RT)	Ref.
B, U (5)	2	50 × 1.8	Zipax® SCX (NA)	E-1[a]	NA	UV (280)	4.5	2-Aminopyridine (12)	a	1

Thin-Layer Chromatography

Specimen (mℓ)	S	Plate (manufacturer)	Layer (mm)	Solvent	Post-separation treatment	Det. (nm)	R_f	Internal standard (R_f)	Other compounds (R_f)	Ref.
B, U (5)	3	20 × 20 cm (Merck)	Silica gel (0.25)	S-1	Sp: 0.1% 2,6-Dibromoquinone-4-chlorimide in methanol	Visual	0.36	—	5′-Desoxy-5′-methylsulfinylpyridoxal (0.3) 5′-Desoxy-5′-methylthiopyridoxal (0.9)	1

Note: E-1 = 0.05 *M* Citrate phosphate buffer, pH 5.65; 0.3 *M* sodium perchlorate + 2% 2-propanol.

S-1 = Acetone + 28% ammonia.
17 : 3

[a] Metabolites are analyzed with a different mobile phase.

REFERENCE

1. Kitao, K., Yiata, N., and Kamada, A., *Chem. Pharm. Bull.*, 25, 1335, 1975.

PYRITHYLDIONE

Gas Chromatography

Specimen (mℓ)	S	Column m × mm	Packing (mesh)	Oven temp (°C)	Gas (mℓ/min)	Det.	RT (min)	Internal standard (RT)	Deriv.	Other compounds (RT)	Ref.
Tissue (5—10 g)	3	1.9 × 4	3% OV-275 Chromosorb® W (NA)	200	NA (37)	FID	4.5	Methyprylon (2.7)	—	Glutethimide (7.9)	1

REFERENCE

1. Martens, F. K., Martens, M. A., Demeter, J., and Heyndrickx, A., *J. Pharm. Sci.*, 65, 1393, 1976.

QUINIDINE

Gas Chromatography

Specimen (mℓ)	S	Column m × mm	Packing (mesh)	Oven temp (°C)	Gas (mℓ/min)	Det.	RT (min)	Internal standard (RT)	Deriv.	Other compounds (RT)	Ref.
B (1)	2	1.2 × 2 Steel	? OV-7 Chromosorb® W (80/100)	270	Nitrogen (20)	FID	6.9	Cinchonidine (4.4)	Methyl	—	1
B (1)	2	1.8 × 2	3% OV-17 Chromosorb® W (80/100)	Temp. Progr.	Nitrogen (14.5)	FID	9.1	Cinchonine (5.9)	Methyl	—	2
B (1)	2	1.8 × ?	3% OV-17 Gas Chrom® Q (80/100)	245	Nitrogen (30)	FID	7.2	Cinchonidine (4.3)	Methyl	—	3
B (1)	2	2.2 × 2	3% OV-1 Gas Chrom® Q (100/120)	270	Nitrogen (35)	NPD	3.6	Chloroquine (2.2)	—	—	4

High-Pressure Liquid Chromatography

Specimen (mℓ)	S	Column cm × mm	Packing (μm)	Elution	Flow rate (mℓ/min)	Det. (nm)	RT (min)	Internal standard (RT)	Other compounds (RT)	Ref.
B (.05)	1	30 × 4	μ-Bondapak® C_{18} (10)	E-1	1.8	Fl (Ex: 340, Fl: KV418 filter	7.1	Cinchonidine (8.5)	3-Hydroxyquinidine (3.0) O-Desmethylquinidine (3.8) 2′-Oxoquinidine (15)	5
B (0.5)	2	N	μ-Bondapak® C_{18} (10)	E-2	2	UV (254)	4.2	Theoluomine (2.2)	Dihydroquinidine (5.5)	6
B (0.2)	2	30 × 4	μ-Bondapak®	E-3	2	UV (33)	2.7		—	7

QUINIDINE (continued)

High-Pressure Liquid Chromatography

Specimen (mℓ)	S	Column cm × mm	Packing (μm)	Elution	Flow rate (mℓ/min)	Det. (nm)	RT (min)	Internal standard (RT)	Other compounds (RT)	Ref.
			Alkyl-phenyl (10)							
B (1)	2	25 × 4.6	Partisil®-10 (10)	E-4	1.2	UV (280)	3	Cinchonidine (3.2)	Dihydroquinidine (4)	8
B (0.3)	2	25 × 4.6	Partisil®-silica (10)	E-5	1.1	UV (236)	6	Strychnine (11)	Dihydroquinidine (7)	9
B (0.5)	1	25 × 3	LiChrosorb® Si 60 (5)	E-6	3.0	Fl (Ex: 325, Fl: 420)	4	—	Dihydroquinidine (3.5)	10
B (0.1)	2	25 × 2.1	MicroPak® MCH (10)	E-7	1.0	UV (254)	7.5	Quinine (8.5)	3-Hydroxyquinidine (3)	11
U (1)	2	30 × 4	μ-Bondapak® Alkyl-phenyl (10)	E-8	1.0	UV (230)	22.5	Oxprenolol (15)	3-Hydroxyquinidine (11.4) 2′-Quinidinone (7.5)	12

Thin-Layer Chromatography

Specimen (mℓ)	S	Plate (manufacturer)	Layer (mm)	Solvent	Post-separation treatment	Det. (nm)	R_f	Internal standard (R_f)	Other compounds (R_f)	Ref.
B (0.01)	2	20 × 20 (Merck)	Silica gel 60 (0.25)	S-1	Sp: 50% Sulfuric acid (8) + ethanol (2)	Fl (Ex: 366, Fl: 456)	NA	—	—	13
B, U (0.5—1)	2	20 × 20 (Merck)	Silica gel 60 (0.25)	S-2	—	Reflectance (278)	0.45	—	Dihydroquinidine (0.26)	14

Note: E-1 = Acetonitrile + 2.5% acetic acid.
12 : 88

E-2 = Methanol + acetic acid + water, pH 2.6.
25 : 4 : 71

E-3 = Acetonitrile + 0.75 *M* sodium acetate buffer, pH 3.6.
40 : 60

E-4 = Methanol + 1 *M* ammonium nitrate + 2 *M* ammonium hydroxide.
27 : 2 : 1

E-5 = 0.75% Ammonium hydroxide in methanol.

E-6 = Dichloromethane + hexane + methanol + 70% perchloride acid.
60 : 35 : 5.5 : 0.1

E-7 = Methanol + 0.01 *M* potassium dihydrogen phosphate + phosphoric acid.
10 : 90 : 0.85

E-8 = 0.05 *M* Phosphate buffer, pH 4.5 + acetonitrile + tetrahydrofuran.
80 : 15 : 5

S-1 = Benzene + dioxane + ethanol + 25% ammonia.
50 : 40 : 5 : 5

S-2 = Methanol + acetone.
5 : 1

REFERENCES

1. **Midha, K. K. and Charette, C.,** *J. Pharm. Sci.,* 63, 1244, 1974.
2. **Valentine, J. L., Dariscoll, P., Hamburg, E. L., and Thompson, E. D.,** *J. Pharm. Sci.,* 65, 96, 1976.
3. **Huffman, D. H. and Hignite, C. E.,** *Clin. Chem.,* 22, 810, 976.
4. **Moulin, M. A. and Kinsun, H.,** *Clin. Chim. Acta,* 75, 491, 1977.
5. **Drayer, D. E., Restivo, K., and Reidenberg, M. M.,** *J. Lab. Clin. Med.,* 90, 816, 1977.
6. **Crouthamel, W. G., Kowarski, B., and Narang, P. K.,** *Clin. Chem.,* 23, 2030, 1977.
7. **Powers, J. L. and Sadee, W.,** *Clin. Chem.,* 24, 299, 1978.
8. **Peat, M. A. and Jennison, T. A.,** *Clin. Chem.,* 24, 2166, 1978.
9. **Achari, R. G., Baldridge, J. L., Koziol, T. R. and Yu, L.,** *J. Chromatogr.,* 16, 271, 1978.
10. **Sved, S., McGilveray, I. J., and Beaudoin, N.,** *J. Chromatogr.,* 145, 437, 1978.

QUINIDINE (continued)

11. Weidner, N., Ladenson, J. H., Larson, L., Kessler, G., and McDonald, J. M., *Clin. Chim. Acta,* 91, 7, 1979.
12. Bonora, M. R., Guentert, T. W., Upton, R. A., and Riegelman, S., *Clin. Chim. Acta,* 91, 277, 1979; *J. Chromatogr.,* 162, 59, 1979.
13. Christiansen, J., *J. Chromatogr.,* 123, 57, 1976.
14. Wesley-Hadzija, B. and Mattocks, A., *J. Chromatogr.,* 144, 223, 1977.

RESERPINE

High-Pressure Liquid Chromatography

Specimen (mℓ)	S	Column cm × mm	Packing (μm)	Elution	Flow rate (mℓ/min)	Det. (nm)	RT (min)	Internal standard (RT)	Other compounds (RT)	Ref.
D	3	25 × 2.1	LiChrosorb® SI60 (5)	E-1	1.5	UV (254)	1.1	Polythiazide (1.8)	Hydrochlorothiazide (5.3)	1
B (2)	1	30 × 4	μ-Bondapak® C_{18} (10)	E-2	2.5	Fl[a] (Ex: 390, Fl: 470)	6	—	—	2

Thin-Layer Chromatography

Specimen (mℓ)	S	Plate (manufacturer)	Layer (mm)	Solvent	Post-separation treatment	Det. (nm)	R_f	Internal standard (R_f)	Other compounds (R_f)	Ref.
B (1—4)	3	10 × 10 cm (Merck)	Silica gel 60[b] (NA)	S-1[c]	E: Acetic acid vapors	Visual (360)	0.5	—	—	3

Note: E-1 = *n*-Hexane + 2-propanol + chloroform + diethylamine.
77 : 18 : 5 : 0.01

E-2 = Methanol + 0.01 *M* sodium heptanesulfonate in water.
65 : 35

S-1 = Chloroform + methanol.
95 : 5

[a] Oxidation with vanadium pentaoxide prior to chromatography.
[b] High performance thin-layer plates.
[c] Alternative solvents are also described.

RESERPINE (continued)

REFERENCES

1. Butterfield, A. G., Lovering, E. G., and Sears, R. W., *J. Pharm. Sci.*, 67, 650, 1978.
2. Sams, R., *Anal. Lett.*, B11, 697, 1978.
3. Sams, R. A. and Huffman, R., *J. Chromatogr.*, 161, 410, 1978.

RIBAVIRIN

Gas Chromatography

Specimen (mℓ)	S	Column m × mm	Packing (mesh)	Oven temp (°C)	Gas (mℓ/min)	Det.	RT (min)	Internal standard (RT)	Deriv.	Other compounds (RT)	Ref.
B (2) U (0.2)	1	1 × 2	3% OV-17 Chromosorb® W (80/100)	Temp. Progr.	Methane (NA)	MS-CI	1.4	Arabinose analog (1.9)	Trimethylsilyl	—	1

REFERENCE

1. Roboz, J. and Suzuki, R., *J. Chromatogr.*, 160, 169, 1978.

RIBOFLAVIN

Thin-Layer Chromatography

Specimen (mℓ)	S	Plate (manufacturer)	Layer (mm)	Solvent	Post-separation treatment	Det. (nm)	R_f	Internal standard (R_f)	Other compounds (R_f)	Ref.
U (NA)	2	NA	Silica gel H (0.2)	S-1	—	Fl (reflectance) (NA)	0.7	—	—	1

Note: S-1 = Pyridine + acetic acid + water.
19 : 2 : 79

REFERENCE

1. Krämer, U., Bitsch, R., and Hötzel, D., *Klin. Wochenschr.*, 55, 243, 1977.

RIFAMPICIN

High-Presure Liquid Chromatography

Specimen (mℓ)	S	Column cm × mm	Packing (μm)	Elution	Flow rate (mℓ/min)	Det. (nm)	RT (min)	Internal standard (RT)	Other compounds (RT)	Ref.
B, U (1) Saliva (1)	1	10 × 7.5	LiChrosorb® SI60 (5)	E-1	3	UV (254)	3	—	3-Formylrifamycin (6) 25-Desacetylrifampicin (7.5) 3-Formyl-25-desacetylrifamycin (9.5)	1
D	3	25 × 2	MicroPak® NH_2 (10)	E-2	0.7	UV (334)	5	—	Rifampicinquinone (2)	2

Thin-Layer Chromatography

Specimen (mℓ)	S	Plate (manufacturer)	Layer (mm)	Solvent	Post-separation treatment	Det. (nm)	R_f	Internal standard (R_f)	Other compounds (R_f)	Ref.
D	3	20 × 20 cm (Merck)	Silica gel 60F (0.25)	S-1	—	Visual	0.6	—	Rifampicinquinone (0.75) 25-Desacetyl rifampicin (0.41) Rifampicin-*N*-oxide (0.09)	3

Note: E-1 = Dichloromethane + isooctane + ethanol + water + acetic acid.
36.6 : 45 : 16.8 : 1.65 : 0.002

E-2 = Chloroform + methanol.
97 : 3

S-1 = Chloroform + methanol + water.
80 : 20 : 2.5

RIFAMPICIN (continued)

REFERENCES

1. Lecaillon, J. B., Febvre, N., Metayer, J. P., and Souppart, C., *J. Chromatogr.*, 145, 319, 1978.
2. Vlasakova, V., Benes, J., and Zivny, K., *J. Chromatogr.*, 151, 199, 1978.
3. Wilson, W. L., Graham, K. C., and Lebelle, M. J., *J. Chromatogr.*, 144, 270, 1977.

Ro 7-9957

Gas Chromatography

Specimen (mℓ)	S	Column m × mm	Packing (mesh)	Oven temp (°C)	Gas (mℓ/min)	Det.	RT (min)	Internal standard (RT)	Deriv.	Other compounds (RT)	Ref.
B (1) U (5)	1	1.2 × 4	3% OV-17 Gas Chrom® Q (60/80)	250	Argon 9 + methane 1 (120)	ECD	8	Diazepam (4.5)	—	*N*-Desmethyl metabolite (11.4)	1

REFERENCE

1. **Puglisi, C. V., de Silva, J. A. F., and Leon, A. S.,** *J. Chromatogr.*, 118, 371, 1976.

ROSOXACIN

High-Pressure Liquid Chromatography

Specimen (mℓ)	S	Column cm × mm	Packing (μm)	Elution	Flow rate (mℓ/min)	Det. (nm)	RT (min)	Internal standard (RT)	Other compounds (RT)	Ref.
B, U (1)	1	2 5 × 4.6	Partisil®-PXS 10/25 (10)	E-1	2	UV (280)	12	a (7 .5)	Rosoxacin-*N*-oxide (5.4)	1

Note: E-1 = Acetonitrile + 0.2 *M* phosphoric acid.
92 : 8

[a] 7-(2,6-Dimethyl-4-pyridyl)-1-ethyl-1,4-dihydro-4-oxo-3-quinoline-carboxylic acid N-oxide.

REFERENCE

1. **Kullberg, M. P., Koss, R., O'Neil, S., and Edelson, J.,** *J. Chomatogr.*, 173, 155, 1979.

SACCHARIN

Gas Chromatography

Specimen (mℓ)	S	Column m × mm	Packing (mesh)	Oven temp (°C)	Gas (mℓ/min)	Det.	RT (min)	Internal standard (RT)	Deriv.	Other compounds (RT)	Ref.
D	3	0.9 × 3 Steel	5% OV-210 Varaport® 30 (100/120)	160	Nitrogen (30)	FID	7	*n*-Octacosane (13)	Trimethyl-silyl	—	1

High-Pressure Liquid Chromatography

Specimen (mℓ)	S	Column cm × mm	Packing (μm)	Elution	Flow rate (mℓ/min)	Det. (nm)	RT (min)	Internal standard (RT)	Other compounds (RT)	Ref.
D	3	30 × 1.9	μ-Bondapak® C_{18} (10)	E-1	1.5	UV (280)	NA[a]	Theophylline (NA)	—	2

Note: E-1 = Methanol + acetic acid + water.
60 : 1 : 139

[a] Saccharin is eluted before theophylline. Both compounds are eluted within 5 min.

REFERENCES

1. **Ratchik, E. M. and Viswanathan, V.**, *J. Pharm. Sci.*, 64, 133, 1975.
2. **Palermo, P. J. and Tsai, P. S.-F.**, *J. Pharm. Sci.*, 68, 878, 1979.

SADDAMINE

Gas Chromatography

Specimen (mℓ)	S	Column m × mm	Packing (mesh)	Oven temp (°C)	Gas (mℓ/min)	Det.	RT (min)	Internal standard (RT)	Deriv.	Other compounds (RT)	Ref.
D	3	2 × 2.5 Steel	5% OV-1 Gas Chrom® Q (100/120)	275	Nitrogen (30)	FID	7.5	—	Trimethyl-silyl	Phenol (1) Benzylamine (2.2) Salicylaldehyde (3.2) Salicylic acid (4.8)	1

REFERENCE

1. Wood, S. G., Al-Ani, M. R., and Lawson, A., *J. Pharm. Sci.*, 68, 374, 1979.

SALBUTAMOL

Gas Chromatography

Specimen (mℓ)	S	Column m × mm	Packing (mesh)	Oven temp (°C)	Gas (mℓ/min)	Det.	RT (min)	Internal standard (RT)	Deriv.	Other compounds (RT)	Ref.
B (2)	1	2.7 × 4	3% OV-210 Gas Chrom® Q (100/120)	Temp.	Helium (40)	MS-EI	11.5	[$^2H_3$]-Salbutamol (11.5)	Trimethylsilyl[a]	—	1

[a] *t*-Butyldimethylsilyl derivative was analyzed on a 3% OV-101 (1 m × 4 mm) column at 250°C and a helium flow of 30 mℓ/min (retention time − 3.5 min).

REFERENCE

1. **Martin, L. E., Rees, J., and Tanner, R. J. N.,** *Biomed. Mass Spectrom.*, 3, 184, 1976.

SALICYLAMIDE

Gas Chromatography

Specimen (mℓ)	S	Column m × mm	Packing (mesh)	Oven temp (°C)	Gas (mℓ/min)	Det.	RT (min)	Internal standard (RT)	Deriv.	Other compounds (RT)	Ref.
B (1)	1	1.8 × 4	5% OV-17 CQ (80/100)	Temp. Progr.	Nitrogen (50)	FID	17	*m*-Toluic acid (5)	Trimethylsilyl	Salicylic acid (8) Acetylsalicylic acid (12)	1
B (1) U (0.05) Saliva (1)	2	1.2 × 0.8	2% OV-225 + 1% OV-17 Gas Chrom® Q (80/100)	200	Helium (10)	NPD	1	*N*-Methylhexobarbital (1.8)	—	—	2

REFERENCES

1. Rance, M. J., Jordan, B. J., and Nichols, J. D., *J. Pharm. Pharmacol.*, 27, 425, 1975.
2. de Boer, A. G., Gubbens-Stibbe, J. M., deKonning, F. H., Bosma, A., and Breimer, D. D., *J. Chromatogr.*, 162, 457, 1979.

SALICYLANILIDE

High-Pressure Liquid Chromatography

Specimen (mℓ)	S	Column cm × mm	Packing (μm)	Elution	Flow rate (mℓ/min)	Det. (nm)	RT (min)	Internal standard (RT)	Other compounds (RT)	Ref.
D	3	25 × 4.6	Partisil® 10 ODS (10)	E-1	1	UV (254)	NA[a]	—	3,4′,5-Tribromo-salicylanilide (NA) 4′,5-Dibromosali-cylanilide (NA) 4′-Monobromo-salicylanilide (NA)	1

Note: E-1 = Methanol + water.
7 : 3

[a] The chromatogram shows the elution order: 1, tribromosalicylanilide; 2, salicylanilide; 3, monobromosalicylanilide; and 4, dibromosalicylanilide.

REFERENCE

1. **Cukor, P., Persiani, C., Woehler, M., and Bischak, N.**, *J. Chromatogr.*, 147, 496, 1978.

SALICYLIC ACID[a]

Gas Chromatography

Specimen (mℓ)	S	Column m × mm	Packing (mesh)	Oven temp (°C)	Gas (mℓ/min)	Det.	RT (min)	Internal standard (RT)	Deriv.	Other compounds (RT)	Ref.
B (NA)	2	1.2 × 4	3% OV-17 Gas Chrom® Q (100/120)	130	Helium (18)	FID	3.5	*p*-Toluic acid (1.8)	Trimethyl-silyl	—	1
D	3	1.8 × 2 Steel	Tenax GC (60/80)	Temp. Progr.	Helium (30)	FID	6	—	—	—	2

High-Pressure Liquid Chromatography

Specimen (mℓ)	S	Column cm × mm	Packing (μm)	Elution	Flow rate (mℓ/min)	Det. (nm)	RT (min)	Internal standard (RT)	Other compounds (RT)	Ref.
B (0.2)	1	15 × 3	5% Trioctylamine on Kieselghur (5—7)	E-1	b	UV (235)	4.2	—	—	3
B (0.1)	1	30 × 4	μ-Bondapak® C_{18} (10)	E-2	1	UV (237)	7	Phthalic acid (3.7)	Salicyluric acid (4.6) Acetylsalicylic acid (5.3)	4
B (0.05)	2	30 × 4	μ-Bondapak® C_{18}[c] (10)	E-3	1.5	UV (248)	2.5	8-Chlorotheo-phylline (5)	Acetaminophen (1.8)	5
B (0.02)	1	30 × 4	μ-Bondapak® C_{18} (10)	E-4	2.6	UV (313)	11.3	*O*-Methoxyben-zoic acid (7.7)	Gentisic acid (3.9) Salicyluric acid (5.4)	6
B (0.003)	2	30 × 4	μ-Bondapak® C_{18} (10)	E-5	2	Fl (Ex: 300, Fl: d)	5	*O*-Methoxyben-zoic acid (3)	Salicylamide (1.2)	7

Thin-Layer Chromatography

Specimen (mℓ)	S	Plate (manufacturer)	Layer (mm)	Solvent	Post-separation treatment	Det. (nm)	R_f	Internal standard (R_f)	Other compounds (R_f)	Ref.
B (1) U (10)	2	NA (Merck)	Silica gel 60[e]	S-1	—	UV refractive mode (325)	0.8	—	Salicyluric acid (0.32) Gentisic acid (0.63)	8
B (0.1)	2	20 × 20 cm (Merck)	Silica gel 60 (0.25)	S-2	Ex: Ammonia vapors (15 min)	Fl Reflectance (Ex: 315, Fl: 405[f])	0.6	—	—	9

Note: E-1 = 0.25 *M* Perchloric acid saturated with trioctylamine, pH 0.7.

E-2 = Acetonitrile + 0.05% phosphoric acid, pH 2.5.
35 : 65

E-3 = 2-Propanol + 0.02 *M* phosphoric acid, pH 2.9.
30 : 970

E-4 = Acetic acid + methanol + water.
41 : 180 : 779

E-5 = Methanol + 20 m*M* nitric acid.
50 : 50

S-1 = Diethyl ether + *n*-butyric acid.
10 : 1

S-2 = Chloroform + ethyl acetate + formic acid.
18 : 33 : 0.6

[a] Acetylsalicylic acid (aspirin) is rapidly changed to salicylic acid in vivo. In emergency, salicylate in blood is routinely measured by Trinder's colorimetric procedure (*Biochem. J.*, 57, 301, 1954).
[b] Pressure = 100 bar.
[c] Column temperature = 40°C.
[d] Wratten 2A emission filter.

SALICYLIC ACID[a] (continued)

[e] High-performance thin-layer plates.
[f] Turner filter #812.

REFERENCES

1. Tischio, J. P., *J. Pharm. Sci.*, 65, 1530, 1976.
2. Ko, P. H., *J. Pharm. Sci.*, 66, 731, 1977.
3. Terweij-Groen, C. P., Vahlkamp, T., and Kraak, J. C., *J. Chromatogr.*, 145,115, 1978.
4. Peng, G. W., Gadalla, M. A. F., Smith, V., Peng, A., and Chiou, W. L., *J. Pharm. Sci.*, 67, 710, 1978.
5. Micelli, J. N., Aravind, M. K., Cohen, S. N., and Done, A. K., *Clin. Chem.*, 25, 1002, 1979.
6. Cham, B. E., Johns, D., Bochner, F., Imhoff, D. M., and Rowland, M., *Clin. Chem.*, 25, 1420, 1979.
7. Blair, D., Rumack, B. H., and Peterson, R. G., *Clin. Chem.*, 24, 1543, 1978.
8. Chrastil, J. and Wilson, J. T., *J. Chromtogr.*, 152, 183, 1978.
9. Gupta, R. N. and Eng, F., Unpublished data.

SCH-12679

(7,8-Dimethoxy-3-methyl-phenyl-2,3,4,5-tetrahydro-1H-3-benzazepine)

Gas Chromatography

Specimen (mℓ)	S	Column m × mm	Packing (mesh)	Oven temp (°C)	Gas (mℓ/min)	Det.	RT (min)	Internal standard (RT)	Deriv.	Other compounds (RT)	Ref.
B (3)	2	1.8 × 4	3% PC3210[a] Gas Chrom® Q (80/100)	200	Nitrogen (45)	FID	7.3	Protriptyline (5.9)	—	3-Desmethyl metabolite[b] (8.5)	1

[a] 50% SE-30 + 50% OV-210.
[b] Conditions for the estimation of 7- and 8-desmethyl metabolites are also described.

REFERENCE

1. Cooper, S. F., Elie, R., Albert, J. M., Gravel, G. B., and Langlois, Y., *J. Chromatogr.*, 163, 47, 1979.

SL 75212

4-(2-Cyclopropylmethoxyethyl)-1-phenoxy-3-isopropyl-aminopropan-2-ol

Gas Chromatography

Specimen (mℓ)	S	Column m × mm	Packing (mesh)	Oven temp (°C)	Gas (mℓ/min)	Det.	RT (min)	Internal standard (RT)	Deriv.	Other compounds (RT)	Ref.
B (0.5—2)	1	2 × 3	3% OV-17 Chromosorb® W (80/100)	210	Argon (NA) + Methane (NA) (50)	ECD	7.3	Propranolol (4)	Heptafluo-robutyryl	—	1

REFERENCE

1. Bianchetti, G., Ganansia, J., and Morselli, P. L., *J. Chromatogr.*, 176, 134, 1979.

SORBITOL

Gas Chromatography

Specimen (mℓ)	S	Column m × mm	Packing (mesh)	Oven temp (°C)	Gas (mℓ/min)	Det.	RT (min)	Internal standard (RT)	Deriv.	Other compounds (RT)	Ref.
D	3	1.2 ×(NA)	2% OV-17 Chromosorb® G (80/100)	220	Helium (55)	FID	6	*m*-Diphenylbenzene (8.5)	Tris-*n*-butylboronate esters	Mannitol (4.5)	1

REFERENCE

1. Sondack, D. L., *J. Pharm. Sci.*, 64, 128, 1975.

SPECTINOMYCIN

Gas Chromatography

Specimen (mℓ)	S	Column m × mm	Packing (mesh)	Oven temp (°C)	Gas (mℓ/min)	Det.	RT (min)	Internal standard (RT)	Deriv.	Other compounds (RT)	Ref.
D	3	0.6 × 3	5% SE-52 Diatoport® S (80/100)	190	Helium (60)	FID	9	Triphenylantimony (5)	Trimethylsilyl	Actinamine (2)	1

High-Pressure Liquid Chromatography

Specimen (mℓ)	S	Column cm × mm	Packing (μm)	Elution	Flow rate (mℓ/min)	Det. (nm)	RT (min)	Internal standard (RT)	Other compounds (RT)	Ref.
D	3	25 × 4.6	LiChrosorb® RP-8 (NA)	E-1	2	Fl[a] (Ex: 350, Fl: 450)	7	—	Actinamine (3)	2

Note: E-1 = 0.02 *M* Sodium heptanesulfonate, 0.2 *M* sodium sulfate, and 0.1% acetic acid in water.

[a] Post-column reaction with *o*-phthalaldehyde.

REFERENCES

1. Brown, L. W. and Bowman, P. B., *J. Chromatogr. Sci.*, 12, 373, 1974.
2. Myers, H. N. and Rindler, J. V., *J. Chromatogr.*, 176, 103, 1979.

SPIRAMYCIN

High-Pressure Liquid Chromatography

Specimen (mℓ)	S	Column cm × mm	Packing (μm)	Elution	Flow rate (mℓ/min)	Det. (nm)	RT (min)	Internal standard (RT)	Other compounds (RT)	Ref.
D	3	10 × 3.2	R Sil-C_{18}[a] (10)	E-1	0.8	UV (232)	1[b] = 6 2[b] = 8 3[b] = 11.5	—	—	1

Note: E-1 = Acetonitrile + water + diethylamine.
50 : 49.95 : 0.05

[a] Column and flow cell temperature = 50°C.
[b] Components of spiramycin preparations.

REFERENCE

1. Bens, G. A., van den Bossche, W., and de Moerloose, P., *Chromatographia*, 12, 294, 1979.

STREPTOZOCIN

High-Pressure Liquid Chromatography

Specimen (mℓ)	S	Column cm × mm	Packing (μm)	Elution	Flow rate (mℓ/min)	Det. (nm)	RT (min)	Internal standard (RT)	Other compounds (RT)	Ref.
D	3	30 × 4	μ-Bondapak® C_{18} (10)	E-1	0.5	UV (254)	α = 7 β = 5	Potassium hydrogen phthalate (21)	—	1

Note: E-1 = 0.1 *M* Acetic acid in water + methanol, pH 4.0.
97 : 3

REFERENCE

1. Oles, P. J., *J. Pharm. Sci.*, 67, 1300, 1978.

STRYCHNINE

High-Pressure Liquid Chromatography

Specimen (mℓ)	S	Column cm × mm	Packing (μm)	Elution	Flow rate (mℓ/min)	Det. (nm)	RT (min)	Internal standard (RT)	Other compounds (RT)	Ref.
D	3	30 × 2	Merckosorb® Si60 (5)	E-1	1.15	UV (254)	11	—	Brucine (18)	1
Stomach wash Grain bait	3	30 × 4	μ-Bondapak® C_{18} (10)	E-2	1.5	UV (254)	3	—	—	2

Note: E-1 = Diethyl ether + methanol.
1 : 1

E-2 = Methanol + 0.005 *M* phosphate buffer, pH 3.0.
40 : 60

REFERENCES

1. **Verpoorte, R. and Svendesen, A. B.,** *J. Chromatogr.*, 100, 231, 1974.
2. **Crouch, M. D. and Short, C. R.,** *J. Assoc. Off. Anal. Chem.*, 61, 612, 1978.

SUCCINYLDICHOLINE BROMIDE

Thin-Layer Chromatography

Specimen (mℓ)	S	Plate (manufacturer)	Layer (mm)	Solvent	Post-separation treatment	Det. (nm)	R_f	Internal standard (R_f)	Other compounds (R_f)	Ref.
B (0.01)[a]	2	20 × 20 cm (Merck)	Cellulose (0.1)	S-1	—	Determination of (C^{14}) radioactivity by radiochromatoscanning	0.14	—	Succinylmonocholine (0.31) Choline (0.54)	1

Note: S-1 = 2-propanol + methanol + water.
5 : 10 : 2

[a] Determination of cholinesterase activity using ^{14}C-labeled succinyldicholine as the substrate.

REFERENCE

1. **Agarwal, D. P. and Goedde, H. W.,** *J. Chromatogr.*, 121, 170, 1976.

SULBENICILLIN

High-Pressure Liquid Chromatography

Specimen (mℓ)	S	Column cm × mm	Packing (μm)	Elution	Flow rate (mℓ/min)	Det. (nm)	RT (min)	Internal standard (RT)	Other compounds (RT)	Ref.
U (0.02)	1	30 × 4	μ-Bondapak® C_{18} (10)	E-1	3.0	UV (254)	23	—	Carbenicillin 5, 7)[a]	1

Note: E-1 = Methanol + 0.01 *M* tetrabutylammonium bromide.
4 : 7

[a] Two unresolved peaks are probably due to the diasteromers.

REFERENCE

1. **Yamaoka, K., Narita, S., Nakagawa, T., and Uno., T.,** *J. Chromatogr.*, 168, 187, 1979.

SULFALSEPINE

Gas Chromatography

Specimen (mℓ)	S	Column m × mm	Packing (mesh)	Oven temp (°C)	Gas (mℓ/min)	Det.	RT (min)	Internal standard (RT)	Deriv.	Other compounds (RT)	Ref.
B (0.2) U (1)	2	1.5 × 4	3% OV-1 Supelcoport (80/100)	250	Nitrogen (90)	ECD	6.5	*N*-Butylchlorthalidone (7.9)	Methyl[a]	Li/Ba 589[b] (8.5)	1

[a] Extractive methylation in the presence of methyl iodide and tetrahexylammonium hydrogen sulfate.
[b] Metabolite of sulfalseprine.

REFERENCE

1. Degen, P. H. and Schweizer, A., *J. Chromatogr.*, 142, 549, 1977.

SULFINPYRAZONE

Gas Chromatography

Specimen (mℓ)	S	Column m × mm	Packing (mesh)	Oven temp (°C)	Gas (mℓ/min)	Det.	RT (min)	Internal standard (RT)	Deriv.	Other compounds (RT)	Ref.
B (NA)	2	1.8 × 2	3% OV-17 Chromosorb® W (80/100)	250	Nitrogen (30)	FID	NA	Phenylbutazone (NA)	Methyl[a]	—	1
B, U (1)	1	1 × 2	2% OV-17 Gas Chrom® Q (NA)	200	Nitrogen (30)	NPD	2.1	Phenylbutazone (3.2)	Methyl[b]	*p*-Hydroxysulfin-pyrazone (5.6) G-31442[c] (7)[d]	2

High-Pressure Liquid Chromatography

Specimen (mℓ)	S	Column cm × mm	Packing (μm)	Elution	Flow rate (mℓ/min)	Det. (nm)	RT (min)	Internal standard (RT)	Other compounds (RT)	Ref.
B (0.5)	1	25 × 2.2	MicroPak® SI-10 (10)	E-1	0.7	UV (254)	5	—	—	3
B (1)	1	12 × 4.7	LiChrosorb® Si60 (5)	E-2	2	UV (254)	3—5	—	*p*-Hydroxysulfin-pyrazone (5) G-31442[c] (1.8)	4
B (0.05-0.1)	1	25 × 2.8	LiChrosorb® RP-2 (10)	E-3	0.5	UV (275)	5	Warfarin (10)	—	5

Note: E-1 = Dioxane + methanol.
65 : 35

E-2 = Dichloromethane + ethanol + water + acetic acid.
79.1 : 19 : 1.9 : 0.002

E-3 = 0.1 *M* Ammonium acetate in acetonitrile + water.
35 : 65

SULFINPYRAZONE (continued)

[a] Extractive alkylation in the presence of tetrahexylammonium hydroxide and methyl iodide.
[b] With methyl iodide in the presence of tetramethylammonium hydroxide.
[c] Sulfone metabolite of sulfinpyrazone.
[d] Retention time at column temperaure of 280°C.

REFERENCES

1. **Rosenfeld, J., Buchanan, M., Powers, P., Hirsh, J., Barnett, H. J. M., and Stuart, R. K.,** *Thromb. Res.*, 12, 247, 1978.
2. **Jakobsen, P. and Pedersen, A. K.,** *J. Chromatogr.*, 163, 259, 1979.
3. **Inaba, T., Besley, M. E., and Chow, E. J.,** *J. Chromatogr.*, 104, 165, 1975.
4. **Lecaillon, J.- B. and Souppart, C.,** *J. Chromatogr.*, 121, 227, 1976.
5. **Wong, L. T., Solomonraj, G., and Thomas, B. H.,** *J. Chromatogr.*, 150, 521, 1978.

SULFONAMIDES

1. Phthalyl sulfathiazole; 2. Sulfabenzamide; 3. Sulfisoxazole; 4. Sulfacetamide; 5. Sulfadimethoxine; 6. Sulfachloropyridazine; 7. Sulfadiazine; 8. Sulfaquinoxaline; 9. Sulfamerazine; 10. Sulfamethoxypyridazine; 11. Sulfathiazole; 12. Sulfamethazine; 13. Sulfapyridine

Gas Chromatography

Specimen (mℓ)	S	Column m × mm	Packing (mesh)	Oven temp (°C)	Gas (mℓ/min)	Det.	RT (min)[a]	Internal standard (RT)	Deriv.	Other compounds (RT)	Ref.
D	3	1 × 3	10% OV-101 Chromosorb® G (80/100)	230	Nitrogen (110)	ECD	3 = 30 5 = 90 9 = 51.2 10 = 60 11 = 36.2	Diethyldithio-carbamic acid 2-benzothiazo-lyl ester (12.5)	Methyl[b]	—	1
B, U (1) Tissue	2	2 × 4	3% OV-17 Gas Chrom® Q (100/120)	280	Nitrogen (30)	FID	—	Chlorthalidone (2.2)	Methyl[c]	Sulpiride (3)	2
B (0.2)	2	1.8 × 4	5% OV-17 Gas Chrom® Q (100/120)	285	Argon 95 + Methane 5 (46)	ECD	7 = 4.9	9-Bromophen-threne (1.3)	Methyl[d]	N^4-Acetylsulfadi-azine (10.7)	3
U (25)	2	1.5 × 4	8% Carbowax® 20 *M* + 2% KOH Chromosorb® W (100/120)	Temp. Progr.	Nitrogen (50)	FID	[e]	—	—	—	4
B (0.1)	2	1.2 × 2	5% OV-17 Gas Chrom® Q (80/100)	290	Argon 95 + Methane 5	ECD	13 = 1.8	Sulfamerazine (3)	Methyl[f]	N^4-Acetylsulfapyr-idine (4)	4

High-Pressure Liquid Chromatography

Specimen (mℓ)	S	Column cm × mm	Packing (μm)	Elution	Flow rate (mℓ/min)	Det. (nm)	RT (min)	Internal standard (RT)	Other compounds (RT)	Ref.
D	3	25 × 3.2	LiChrospher®[g] SI 100	E-1	2.5	UV (254)	1 = 3	—	—	6

SULFONAMIDES (continued)

High-Pressure Liquid Chromatography

Specimen (mℓ)	S	Column cm × mm	Packing (μm)	Elution	Flow rate (mℓ/min)	Det. (nm)	RT (min)	Internal standard (RT)	Other compounds (RT)	Ref.
			(10)				2 = 6 3 = 7 4 = 44 5 = 15.5 6 = 8 7 = 36 8 = 9 9 = 37 10 = 26.5 11 = 22 12 = 29 13 = 86			
U (2)	2	50 × 2.6	Amino-Sil-X-1 (NA)	E-2	1.5	UV (254)	9 = 3.8 11 = 9 12 = 2.6	—	*N*-Acetylsulfamethazine (5.8) *N*-Acetylfsulfamerazine (11) *N*-Acetylsulfathiazole (15.6)	7
Swine feed[h]	2	30 × 4	μ-Bondapak® C_{18} (10)	E-3	2	UV (254)	12 = 6.5	—	—	8
B (0.2)	1	30 × 3.9	μ-Bondapak® C_{18} (10)	E-4	1.5	UV (254)	7 = 4.8 9 = 6.4 12 = 8.4	Sulfamethizole (9.2)	—	9
B (0.02)	2	30 × 4	μ-Bondapak® CN (10)	E-5	2.2	UV (254)	13 = 2.8	Sulfamethoxazole (4.7)	*N*-Acetylsulfapyridine (4.0)	10
B (0.1)	2	25 × 4.6	Nucleosil RP-18 (10)	E-6	1.4	UV (254)	13 = 6	Sulfadimidine	N^4-Acetylsulfa-	11

								(10)	pyridine (12) 5-Hydroxysulfa-pyridine (7)	
B (NA)	2	25 × 4	LiChrosorb® Si 100 (10)	E-7	0.6	UV (254)	1 = 3.2 2 = 5.4 3 = 3.5 4 = 3.4 11 = 6.2	—	Sulfamethoxazole (4.8) N⁴-Acetylsulfa-methoxazole (4.1)	12
B (0.5)	2	60 × 2	LiChrosorb® Si 60 (5)	E-8	1	UV (280)	13 = 5.6	Sulfamethazine (7.2)	N⁴-Acetylsulfa-pyridine (11.7)	13
B (1)[c]	1	30 × 4	μ-Bondapak® C_{18} (5)	E-9	0.6	UV (254)	—	Sulfamethoxa-zole (6.2)	Sulfamethizole (4.7)	14

Thin-Layer Chromatography

Specimen (mℓ)	S	Plate (manufacturer)	Layer (mm)	Solvent	Post-separation treatment	Det. (nm)	R_f	Internal standard (R_f)	Other compounds (R_f)	Ref.
B (NA) Tissue Milk	2	20 × 20 cm (Merck)	Silica gel 60 (NA)	S-1	D: i. Fluorescamine (25 mg) in acetone (250 mℓ) ii. 0.5% Triethanola-mine in chloroform	Fl (Ex: 290, Fl: m)	7 = 0.4	—	—	15
B (0.05)	2	10 × 20 cm (Merck)	Silica gel 60 (NA)	S-2	Sp.: Fluorescamine (15 mg) in acetone (200 mℓ)	Fl (Ex: 390, Fl: 490)	—	—	Mafenide (NA)	16
B (5)	2	20 × 20 cm (Laboratory)	Silica gel (0.25)	S-3	D: Fluorescamine (10 mg) in acetone (100 mℓ)	Fl (Ex: 310, Fl: n)	12 = 0.49	Sulfaguanidine (NA)	—	17
D	3	NA (Laboratory)	Silica gel + 1% cadmium ace-tate (0.5)	S-4	Sp: *p*-Dimethylami-nobenzaldehyde in 95% ethanol (100 mℓ) and conc. sulfuric acid (1 mℓ)	Visual	2 = 0.45 7 = 0.18 9 = 0.29 11 = 0.14 13 = 0.34	—	—	18

SULFONAMIDES (continued)

Note: E-1 = 1-Butanol + heptane (saturated with stationary phase).
25 : 75

E-2 = Methanol.

E-3 = Acetonitrile + 1% acetic acid.
30 : 250

E-4 = Acetonitrile + 1% acetic acid.
13 : 87

E-5 = 4% Acetic acid containing 0.4% sodium acetate.

E-6 = Methanol + 1% acetic acid.
20 : 80

E-7 = Methanol + water.
30 : 70

E-8 = Chloroform + acetonitrile + methanol + 35% ammonia.
65.5 : 30 : 4 : 0.5

E-9 = Methanol + water.
15 : 85

S-1 = Diethyl ether (×2 development up to 10–12 cm).

S-2 = Ethyl acetate + methanol + ammonia.
75 : 20 : 5

S-3 = Ethyl acetate.

[a] Numbers before the retention times refer to different sulfonamides. See above for identification.
[b] With dimethylacetal in dimethyl formamide.
[c] Flash methylation with trimethylbenzyl ammonium hydroxide.
[d] With diazomethane.
[e] Pyrolysis-gas chromatography profiles of sulfonamides are described.
[f] Extractive alkylation in the presence of methyl iodide and tetrabutylammonium carbonate.
[g] The column was coated with a stationary phase of 0.3 *M* tetrabutylammonium hydrogen sulphate in 0.1 *M* phosphate buffer, pH 6.8.

[h] Extracts of feed samples were purified by column chromatography prior to analysis by HPLC.
[i] Unbound drug fractions were isolated by Sephadex® gel chromatography.
[m] Total emission above 400 nm.
[n] Total emission above 520 nm.

REFERENCES

1. Nose, N., Kobayashi, S., Hirose, A., and Watanabe, A., *J. Chromatogr.*, 123, 167, 1976.
2. Frigerio, A. and Panarotto, C., *J. Chromatogr.*, 130, 361, 1977.
3. Bye, A. and Land, G., *J. Chromatogr.*, 139, 181, 1977.
4. Irwin, W. J. and Slack, J. A., *J. Chromatogr.*, 153, 526, 1978.
5. Gyllenhall, O., Näslund, B., and Hartvig, P., *J. Chromatogr.*, 156, 330, 1978.
6. Su,S. C., Hartkopf, A. V., and Karger, B. L., *J. Chromatogr.*, 119, 523, 1976.
7. Sharma, J. P., Perkins, E. G., and Bevill, R. F., *J. Pharm. Sci.*, 65, 1606, 1976.
8. Allred, M. C. and Dunmire, D. L., *J. Chromatogr. Sci.*, 16, 533, 1978.
9. Goehl, T. J., Mathur, L. K., Strum, J. D., Jaffe, J. M., Pitlick, W. H., Shah, V. P., Poust, R. I., and Colaizzi, J. L., *J. Pharm. Sci.*, 67, 404, 1978.
10. Owerbach, J., Johnson, N. F., Bates, T. R., Pieniaszek, H. J., Jr., and Jusko, W. J., *J. Pharm. Sci.*, 67, 1250, 1978.
11. Fischer, C. and Klotz, U., *J. Chromatogr.*, 146, 157, 1978; ibid, 162, 237, 1979.
12. Harzer, K., *J. Chromatogr.*, 155, 399, 1978.
13. Lanbeck, K. and Lindström, B., *J. Chromatogr.*, 154, 321, 1978; ibid, 174, 490, 1979.
14. Slavesen, B. and Laake, J., *Medd. Nor. Farm. Selsk.*, 41, 29, 1979.
15. Sigel, C. W., Woolley, J. L., Jr., and Nichol, C. A., *J. Pharm. Sci.*, 64, 973, 1975.
16. Steyn, J. M., *J. Chromatogr.*, 143, 210, 1977.
17. Bevill, R. F., Schemske, K. M., Luther, H. G., Dzierzak, E. A., Limpoka, M., and Felt, D. R., *J. Agric. Food. Chem.*, 26, 1201, 1978.
18. Srivestava, S. P., Dua, V. K., Mehrotra, R. N., and Saxena, R. C., *J. Chromatogr.*, 176, 145, 1979.

SULINDAC[a]

High-Pressure Liquid Chromatography

Specimen (mℓ)	S	Column cm × mm	Packing (μm)	Elution	Flow rate (mℓ/min)	Det. (nm)	RT (min)	Internal standard (RT)	Other compounds (RT)	Ref.
B (0.2)	2	30 × 3.9	μ-Bondapak® C_{18} (10)	E-1	1.5	UV (254)	2.3	Indomethacin (4.7)	Sulindac sulfone (3.3) Sulindac sulfide (7.6)	2

Note: E-1 = Acetonitrile + 45 m*M* KH_2PO_4, pH 3.
60 : 40

[a] A mass spectrometric procedure for the estimation of sulindac and its metabolites in plasma without prior separation by chromatography has recently been published (Reference 1).

REFERENCES

1. Walker, R. W., Gruber, V. F., Rosenberg, A., Wolf, F. J., and Van den Heuvel, W. J. A., *Anal. Biochem.*, 95, 579, 1979.
2. Dusci, L. J. and Hackett, L. P., *J. Chromatogr.*, 171, 490, 1979.

SULTHIAME

Gas Chromatography

Specimen (mℓ)	S	Column m × mm	Packing (mesh)	Oven temp (°C)	Gas (mℓ/min)	Det.	RT (min)	Internal standard (RT)	Deriv.	Other compounds (RT)	Ref.
B (2)	2	1.8 × 4	3% OV-17 Gas Chrom® Q (80/100)	290	Nitrogen (60)	FID	5.3	Thioridazine (8)	Methyl[a]	—	1

High-Pressure Liquid Chromatography

Specimen (mℓ)	S	Column cm × mm	Packing (μm)	Elution	Flow rate (mℓ/min)	Det. (nm)	RT (min)	Internal standard (RT)	Other compounds (RT)	Ref.
B (0.1)	2	30 × 4	μ-Bondapak®-C_{18} (10)	E-1	1.0	UV (235)	4	—	—	2
B (0.5)	2	10 × 5.5	LiChrosorb® RP-8 (5)	E-2	1.0	UV (248)	2.2	5-Ethyl-5-*p*-tolybarbituric acid (5.5)	—	3

Note: E-1 = Acetonitrile + 45 m*M* KH_2PO_4, pH 3.
45 : 55

E-2 = Acetonitrile + 0.5 *M* acetate buffer, pH 5.8.
30 : 70

[a] With 1 *M* trimethylanilinium hydroxide.

REFERENCES

1. Hackett, L. P. and Dusci, L. J., *Clin. Chim. Acta*, 66, 443, 1976.
2. Hackett, L. P. and Dusci, L. J., *Clin. Toxicol*, 13, 551, 1978.
3. Berry, D. J., Clarke, L. A., and Vallins, G. E., *J. Chromatogr.*, 171, 363, 1979.

SYNEPHRINE

Thin-Layer Chromatography

Specimen (mℓ)	S	Plate (manufacturer)	Layer (mm)	Solvent	Post-separation treatment	Det. (nm)	R_f	Internal standard (R_f)	Other compounds (R_f)	Ref.
D	3	(Whatman)	Filter paper No. 1	S-1	Sp: Diazotized *p*-nitro-sulfanilic acid	Visual[a]	0.63	—	—	1

Note: S-1 = Butanol + ethanol + water + ammonia.
9 : 4 : 2 : 0.2

[a] Spot area was measred planimetrically.

REFERENCE

1. El-Nimar, A. E. M., Ammar, H. O., and Salama, H. A., *Pharmazie,* 32, 402, 1977.

TAMOXIFEN

Gas Chromatography

Specimen (mℓ)	S	Column m × mm	Packing (mesh)	Oven temp (°C)	Gas (mℓ/min)	Det.	RT (min)	Internal standard (RT)	Deriv.	Other compounds (RT)	Ref.
B (1)	2	3 × 3.5	1% OV-1 Gas Chrom® Q (100/120)	NA	NA	MS-EI[a]	NA	b (NA)	—	—	1

High-Pressure Liquid Chromatography

Specimen (mℓ)	S	Column cm × mm	Packing (μm)	Elution	Flow rat (mℓ/min)	Det. (nm)	RT (min)	Internal standard (RT)	Oher compounds (RT)	Ref.
B (NA)	1	30 × 4	μ-Bonapak®-CN (10)	E-1	2	Fl[c] (Ex: 256, Fl: 320)	4	—	4-Hydroxytamox-ifen (7)	2

Note: E-1 = Ethylene chloride + acetonitrile + formic acid + water.
95 : 5 : 0.2 : 0.05

[a] Plasma extract is purifed by chromatography on Lipidex® 5000 (4 × 0.5 cm) columns.
[b] Trans-1-(*p*-β-Dimethylaminoethoxyphenyl)-1,2-diphenylprop-1-ene.
[c] Tamoxifen and its metabolite are converted to fluorescence products by irradiation with UV light prior to chromatography.

REFERENCES

1. **Gaskell, S. J., Daniel, C. P., and Nicholson, R. I.,** *J. Endocrinol.*, 78, 293, 1978; 83, 401, 1979.
2. **Mendenhall, D. W., Kobayashi, H., Shih, F. M. L., Sternson, L. A., Higuchi, T., and Fabian, C.,** *Clin. Chem.*, 24, 1518, 1978.

TANDAMINE

Gas Chromatography

Specimen (mℓ)	S	Column m × mm	Packing (mesh)	Oven temp (°C)	Gas (mℓ/min)	Det.	RT (min)	Internal standard (RT)	Deriv.	Other compounds (RT)	Ref.
B (4)	1	2.7 × NA	3% OV-17 Chromosorb® W (100/120)	270	Nitrogen (30)	NPD	NA	Desmethylimipramine	—	—	1

REFERENCE

1. Gifford, L. A. and Witts, D. J., *Br. J. Clin. Pharmacol.*, 3, 337, 1976.

TERBUTALINE

Gas Chromatography

Specimen (mℓ)	S	Column m × mm	Packing (mesh)	Oven temp (°C)	Gas (mℓ/min)	Det.	RT (min)	Internal standard (RT)	Deriv.	Other compounds (RT)	Ref.
B (1) U (0.1)	1	1.2 × 3	3% OV-1 Gas Chrom® Q (80/100)	165	Methane[a] (24)	MS-CI	NA	Bis(2-ethyl-hexyl) hydrogen phosphate (NA)	Trimethyl-silyl	—	1

Thin-Layer Chromatography

Specimen (mℓ)	S	Plate (manufacturer)	Layer (mm)	Solvent	Post-separation treatment	Det. (nm)	R_f	Internal standard (R_f)	Other compounds (R_f)	Ref.
B (5)	1	20 × 20 cm (Whatman)	Silica gel (0.25)	S-1[b]	Sp: i. 5% Glyoxylic acid in ethanol ii. Heat at 70°C for 30 min.	Fl Reflectance (Ex: 316—400, Fl: >490)	NA	Metaproterenol (NA)	—	2

Note: S-1 = Acetone + formic acid.
100 : 2

[a] Methane is also the reagent gas for CI mass spectrometry.
[b] The plates were developed in the dark.

REFERENCES

1. Leferink, J. G., Wagemaker-Engels, I., Maes, R. A. A., Lamont, H., Pauwels, R., and van der Straeten, M., *J. Chromatogr.*, 143, 299, 1977; *Anal. Toxicol.*, 2, 86, 1978.
2. Tripp, S. L., Williams, E., Roth, W. J., Wagner, W. E., Jr., and Lukas, G., *Anal. Lett.*, B11, 727, 1978.

TETRACYCLINES

1. Anhydrotetracycline; 2. Chlortetracycline; 3. Democlocycline; 4. Doxycycline; 5. Epianhydrotetracycline; 6. Epitetracycline; 7. Methacycline; 8. Minocycline; 9. Oxytetracycline; 10. Rolitetracycline; 11. Tetracycline

High-Pressure Liquid Chromatography

Specimen (mℓ)	S	Column cm × mm	Packing (μm)	Elution	Flow rate (mℓ/min)	Det. (nm)	RT[a] (min)	Internal standard (RT)	Other compounds (RT)	Ref.
D	3	225 × 1.8	Pellionex CP-128 (NA)	E-1	1	UV (254)	6 = 3.7 10 = 6.6 11 = 3.1	—	—	1
D	3	12.5 × 5	Silica SC-TAS (18)	E-2	b	UV (280)	1 = 8 2 = 4.5 5 = 6 6 = 2.8 11 = 3	—	—	2
D	3	100 × 2.1	Zipax®[c] SCX (NA)	E-3	0.5	Absorbance (429)	1 = 8 5 = 5.5 6 = 14.5 11 = 18	—	—	3
D	3	30 × 4.6	μ-Bondapak® C_{18} (10)	E-4 (Gradient)	1	UV (280)	1 = 14 2 = 12 5 = 14 6 = 9.5 11 = 10	—	—	4
B, U (2)	2	30 × 4.5	μ-Bondapak® C_{18} (10)	E-5	1	Absorbance (355)	2 = 7 9 = 4.5 11 = 5	—	—	5

High-Pressure Liquid Chromatography

Specimen (mℓ)	S	Column cm × mm	Packing (μm)	Elution	Flow rate (mℓ/min)	Det. (nm)	RT (min)	Internal standard (RT)	Other compounds (RT)	Ref.
D	3	25 × 4.1	Vydac® TP201 (10)	E-6	1.5	UV (254)	3 = 4 6 = 2 7 = 4.5 8 = 12 9 = 2.5 11 = 3	—	—	6
B (0.5) U (0.1)	2	10 × 2	LiChrosorb® RP8 (5)	E-7	0.5	Absorbance (350)	4 = 6	Demeclocycline (3)	—	7

Thin-Layer Chromatography

Specimen (mℓ)	S	Plate (manufacturer)	Layer (mm)	Solvent	Post-separation treatment	Det. (nm)	R_f	Internal standard (R_f)	Other compounds (R_f)	Ref.
D	3	20 × 20 cm (Laboratory)	Silica gel + Na_2 EDTA[d] (10%, pH 8.5) (0.25—0.3)	S-1	E: Ammona vapors	Fl (Ex: 365, Fl: 514—533)	1 = 0.98 2 = 0.21 5 = 0.16 6 = 0.04 11 = 0.1	—	—	8
D	3	20 × 20 cm (Laboratory)	Cellulose[e] (0.3)	S-2	Sp: i. 0.1 M $MgCl_2$ in 95% ethanol ii. Methanolic triethanolamine (10% v/v)	Fl (Ex: 365, Fl: 540)	2 = 0.72 4 = 0.11 0.66[f] 7 = 0.2 9 = 0.47 11 = 0.35 5	—	—	9

TETRACYCLINES (continued)

Note: E-1 = Ethanol + (0.1 *M* Na^+ + 0.003 *M* $EDTA^{-2}$), pH 4.35.
40 : 60

E-2 = Water + acetonitrile (containing 0.1 *M* overall $HClO_4$).
3 : 1

E-3 = 0.03 *M* EDTA, pH 7.

E-4 = i. Acetonitrile + water + 0.2 *M* phosphate buffer, pH 2.5.
100 : 800 : 100

ii. Acetonitrile + water + 0.2 *M* phosphate buffer, pH 2.5.
600 : 300 : 100

E-5 = 0.01 *M* phosphate buffer, pH 2.4, in water + acetonitrile.
70 : 30

E-6 = 0.1 *M* $(NH_4)_2$EDTA + 1 *M* diethanolamine, pH 7.3 + isopropanol + water.
1 : 50 : 8 : 41

E-7 = Acetonitrile + 0.1 *M* citric acid.
24 : 76

S-1 = Ethylene glycol + water + acetone + ethyl acetate.
2 : 2 : 15 : 15

S-2 = Ethyl acetate saturated with water.

[a] See above for the identification of the tetracycline according to the number.
[b] 1.3 mm/sec.
[c] Column temperature = 35°C.
[d] Ethylenediaminetetraacetate.
[e] 0.2 *M* Na_2HPO_4 (70 mℓ) + 0.1 *M* citric acid (84 mℓ) added to 30 g of cellulose. The plates were then dipped in methanolic ethylene glycol solution (20% V/V).
[f] R_f values of β- and α-doxycyclines, respectively.

REFERENCES

1. Butterfield, A. G., Hughes, D. W., Wilson, W. L., and Pound, N. J., *J. Pharm. Sci.*, 64, 316, 1975.
2. Knox, J. H. and Jurand, J., *J. Chromatogr.*, 110, 103, 1975.
3. Lindauer, R. F., Cohen, D. M., and Munnelly, K. P., *Anal. Chem.*, 48, 1731, 1976.
4. Tsuji, K. and Robertson, J. H., *J. Pharm. Sci.*, 65, 400, 1976.
5. Sharma, J. P., Perkins, E. G., and Bevill, R. F., *J. Chromatogr.*, 134, 441, 1977.
6. Mack, G. D. and Ashworth, R. B., *J. Chromatogr. Sci.*, 16, 93, 1978.
7. de Leenheer, A. P. and Nelis, H. J. C. F., *J. Pharm. Sci.*, 68, 999, 1979.
8. Willekens, G. J., *J. Pharm. Sci.*, 64, 1681, 1975.
9. Szabo, A., Nagy, M. K., and Tömörkeny, E., *J. Chromatogr.*, 151, 256, 1978.

TETRAMISOLE

High-Pressure Liquid Chromatography

Specimen (mℓ)	S	Column cm × mm	Packing (μm)	Elution	Flow rate (mℓ/min)	Det. (nm)	RT (min)	Internal standard (RT)	Other compounds (RT)	Ref.
D	3	15 × 4.7	LiChrosorb® RP-8 (10)	E-1	1.6	UV (254)	2.4	Phenol (6)	—	1

Note: E-1 = 1% Sulfuric acid in water + acetonitrile.
80 : 20

REFERENCE

1. Mourot, D., Delepine, B., Boisseau, J., and Gayot, G., *J. Pharm. Sci.*, 68, 796, 1979.

THEOPHYLLINE

Gas Chromatography

Specimen (mℓ)	S	Column m × mm	Packing (mesh)	Oven temp (°C)	Gas (mℓ/min)	Det.	RT (min)	Internal standard (RT)	Deriv.	Other compounds (RT)	Ref.
B (1)	2	1 × 4	3% OV-17 Gas Chrom® Q (100/120)	210	Argon + 5% methane (60)	ECD	1.8	3-Isobutyl-1-methyl-xanthine (2.3)	Pentafluorobenzoyl	—	1
B (0.02)	2	0.9 × 2	3% OV-17 Gas Chrom® Q (100/120)	Temp. Progr.	Nitrogen (35)	NPD	3.5	3-Isobutyl-1-methyl-xanthine (4.4)	Pentyl	Caffeine (2.1) Theobromine (4.2)	2
B (0.1)	2	1.8 × 2	3% OV-17 Chromosorb® G (100/120)	250	Nitrogen (50)	FID	6.6	Amobarbital (2.9)	Butyl	Caffeine (1.8) Theobromine (7.5)	3
B (0.05)	2	1.8 × 2	3% OV-17 Gas Chrom® Q (100/120)	240	Helium (40)	NPD	1.5	3-Isobutyl-1-methyl-xanthine (2.5)	Pentyl	Caffeine (0.8)	4
B (1)	2	0.9 × 0.8	1% OV-17 + 2% OV-225 Gas Chrom® Q (80/100)	230	Helium (5)	NPD	2.2	Cyheptamide (4.1)	—	—	5
B (0.2)	2	1.8 × 2	3% OV-17 Chrom W-HP (100/120)	Temp. Progr.	Nitrogen (30)	MS-EI	8.2	3-Isobutyl-1-methyl-xanthine (10)	Butyl	Caffeine (6.7)	6
B (0.025)	2	1.8 × 2	3% OV-17 Gas Chrom® Q (100/120)	240	Helium (20)	NPD	1.9	3-Isobutyl-1-methyl-xanthine (2.5)	Pentyl	Caffeine (0.9) Theobromine (2.2)	7
B (0.5)	2	1 × 4	3% HI-EFF-8BP Diatomite CLQ (NA)	255	Argon (45)	NPD	12.0	Heptabarbitone (8.8)	—	Caffeine (2.6) Theobromine (5.4)	8
B (1)	2	2 × 4	2% SP2510-DA Supelcoport (100/120)	245	Nitrogen (50)	FID	3.7	3-Isobutyl-1-methyl-xanthine (4.2)	—	—	9
B (0.1)	2	1.8 × 2	3% OV-17 +	230	Nitrogen	NPD	4.3	3-Isobutyl-1-	—	—	10

THEOPHYLLINE (continued)

Gas Chromatography

Specimen (mℓ)	S	Column m × mm	Packing (mesh)	Oven temp (°C)	Gas (mℓ/min)	Det.	RT (min)	Internal standard (RT)	Deriv.	Other compounds (RT)	Ref.
			0.1% terphthalic acid Chromosorb® 750 (100/120)		(30)			methylxanthine (5.6)			

High-Pressure Liquid Chromatography

Specimen (mℓ)	S	Column cm × mm	Packing (μm)	Elution	Flow rat (mℓ/min)	Det. (nm)	RT (min)	Internal standard (RT)	Other compounds (RT)	Ref.
B (1)	2	30 × 4	μ-Bondapak® C_{18} (10)	E-1	2	UV (254)	5.5	—	—	11
B (0.2)	2	25 × 2.1	Zorbax Sil® (6-8)	E-2	0.4	UV (254)	5	β-Hydroxypropyl-theophylline (8)	Caffeine (4.5)	12
B (0.05)	2	25 × 2.6	ODS-Sil-x-1[a] (13)	E-3	1.5	UV (273)	0.6[b]	8-Chlorotheophylline (1)[b]	—	13
B (0.25)	2	25 × 2.6	Sphinsorb ODS (10)	E-4	NA	UV (280)	2.3	8-Chlorotheophylline (3.8)	—	14
B (0.05)	2	30 × 4	μ-Bondapak® C_{18} (10)	E-5	1.7	UV (280)	5	8-Chlorotheophylline (7)	—	15
	2	30 × 4	μ-Bondapak® C_{18} (10)	E-6	2.0	UV (254)	6	β-Hydroxyethyl-theophylline (8)	Caffeine (11)	16
B (1)	2	30 × 2.9	μ-Bondapak® C_{18} (10)	E-7	1.1	UV (254)	26	Dyphylline (30)	Theobromine (16)	17
B (0.1)	2	30 × 4	μ-Bondapak® C_{18} (10)	E-8	1.5	UV (254)	5	β-Hydroxyethyl-theophylline (6)	—	18
NA	—	25 × 3.2	LiChrosorb® C_8 (10)	E-9	1.0	Electrochemical	5	8-Chlorotheophylline (10)	—	19

B (1)	2	20× 20 (Merck)	Silica gel 60 (0.25)	S-1	—	UV-Reflectance (275)	0.40	—	Caffeine (0.05)	20
B (0.5)	2	20 × 20 (Merck)	Silica gel-F_{254} (0.25)	S-2	—	UV-Reflectance (275)	0.23	3-Isobutyl-1-methyl-xanthine (0.34)	Caffeine (0.44)	21
B (3)	3	20 × 20 Plastic (Merck)	Silica gel-F_{254} (NA)	S-3	—	UV-Absorbance (275)[d]	0.43	—	Caffeine (0.75)	22
B (0.01)	2	20 × 20 (Merck)	Silica gel 60 F_{254} (0.25)	S-4	—	UV-Reflectance (273)	0.30	—	Caffeine (0.44)	23

Note: E-1 = Acetonitrile + 0.01 *M* sodium acetate, pH 4.0.
1 : 9

E-2 = Ethanol + chloroform + heptane + acetic acid.
6 : 56.4 : 37.6 : .075

E-3 = Acetonitrile + water + 1% acetic acid.
10 : 480 : 10

E-4 = Methanol + 0.01 *M* sodium acetate, pH 4.0.
25 : 75

E-5 = Methanol + 1% propionic acid, pH 5.
20 : 80

E-6 = Acetonitrile + 0.01 *M* sodium acetate, pH 4.0.
7 : 93

E-7 = Methanol + 0.05 *M* potassium dihydrogen phosphate, pH 4.7.
12 : 88

E-8 = Acetonitrile + 0.026 *M* sodium acetate, pH 4.0.
1 : 9

E-9 = Ethanol + sodium acetate, pH 4.0.
8 : 92

THEOPHYLLINE (continued)

S-1 = Chloroform + methanol.
90 : 10

S-2 = Chloroform + methanol.
95 : 5

S-3 = Ethyl acetate + methanol + ammonia.
80 : 20 : 15

S-4 = Chloroform + methanol.
90 : 10

[a] Column temp. = 55°C.
[b] Relative retention time. Absolute retention time for 8-chlorotheophylline not available.
[c] Column temp. = 50°C.
[d] Spot cut and eluted and absorbance of elute measured spectroscopically.

REFERENCES

1. **Schwertner, H. A., Ludden, T. M., and Wallace, J. E.,** *Anal. Chem.,* 48, 1875, 1976.
2. **Least, C. J., Jr., Johnson, G. F., and Solomon, H. M.,** *Clin. Chem.,* 22, 765, 1976.
3. **Perrier, D. and Lear, E.,** *Clin. Chem.,* 22, 898, 1976.
4. **Lowry, J. D., Williamson, L. J., and Raisys, V. A.,** *J. Chromatogr.,* 143, 83, 1977.
5. **De Boer, A. G. and Breimer, D. D.,** *Pharm. Weekb.,* 112, 764, 1977.
6. **Sheehan, M., Hertel, R. H., and Kelly, C. T.,** *Clin. Chem.,* 23, 64, 1977.
7. **Joern, W. A.,** *Clin. Chem.,* 24, 1458, 1978.
8. **Chambers, R. E.,** *J. Chromatogr.,* 171, 473, 1979.
9. **Schwertner, H. A.,** *Clin. Chem.,* 25, 212, 1979.
10. **Vinet, B. and Zizian, L.,** *Clin. Chem.,* 25, 156, 1979.
11. **Franconi, L. C., Hawek, G. L., Sandman, B. J., and Haney, W. G.,** *Anal. Chem.,* 48, 373, 1976.
12. **Evensen, M. A. and Warren, B. L.,** *Clin. Chem.,* 22, 851, 1976.

13. Adams, R. F., Vandemark, F. L., and Schmidt, G. J., *Clin. Chem.*, 22, 1903, 1976.
14. Peat, M. A. and Jennison, T. A., *J. Anal. Toxicol.*, 1, 204, 1977.
15. Hill, R. E., *J. Chromatogr.*, 135, 419, 1977.
16. Orcutt, J. J., Kozak, P. P., Jr., Gillman, S. A., and Cummins, L. H., *Clin. Chem.*, 23, 599, 1977.
17. Desiraju, R. K., Sugita, E. T., and Mayock, R. L., *J. Chromatogr. Sci.*, 15, 563, 1977.
18. Shihabi, Z. K., *Clin. Chem.*, 24, 1630, 1978.
19. Greenberg, M. S. and Mayer, W. J., *J. Chromatogr.*, 169, 321, 1979.
20. Wesely-Hadzija, B. and Mattock, A. M., *J. Chromatgr.*, 115, 501, 1975.
21. Gupta, R. N., Eng, F., and Stefanec, M., *Clin. Biochem.*, 11, 42, 1978.
22. Roseboom, H., Lingeman, H., and Wiese, G., *Fresenius Z. Anal. Chem.*, 292, 239, 1978.
23. Riechert, M., *J. Chromatogr.*, 146, 175, 1978.

THIAMINE

High-Pressure Liquid Chromatography

Specimen (mℓ)	S	Column cm × mm	Packing (μm)	Elution	Flow rate (mℓ/min)	Det. (nm)	RT (min)	Internal standard (RT)	Other compound (RT)	Ref.
U (6)	2	25 × 3.2	LiChrosorb® (5)	E-1	0.85	Fl[a] (Ex: 360, Fl: 400)	5	—	—	1
D	2	25 × 4.6	LiChrosorb®-NH_2 (NA)	E-2	2	Fl[b] (Ex: 375, Fl: 430)	1.6	—	Thiamine-monophosphate (3.2) Thiamine-pyrophosphate (4.9) Thiamine-triphosphate (7.2)	2

Note: E-1 = Methanol + diethyl ether.
22 : 88

E-2 = Acetonitrile + 90 m*M* potassium phosphate buffer, pH 8.4.
60 : 40

[a] Thiamine is isolated from urine by ion exchange and treated with alkaline potassium ferricyanide prior to analysis by liquid chromatography.
[b] The samples are oxidized with BrCN prior to chromatography.

REFERENCES

1. Roser, R. L., Andrist, A. H., Harrington, W. H., Naito, H. K., and Lonsdale, H., *J. Chromatogr.*, 146, 43, 1978.
2. Ishii, K., Sarai, K., Sanemori, H., and Kawasaki, T., *Anal. Biochem.*, 97, 191, 1979.

THIAZIDES[a]

1. Bemetizide; 2. Bendrofluazide; 3. Benzthiazide; 4. Chlorothiazide; 5. Cyclopenthiazide; 6. Cyclothiazide; 7. Hydrochlorothiazide; 8. Hydroflumethiazide

High-Pressure Liquid Chromatography

Specimen (mℓ)	S	Column cm× mm	Packing (μm)	Elution	Flow rate (mℓ/min)	Det. (nm)	RT[b] (min)	Internal standard (RT)	Other compounds (RT)	Ref.
B (5)	1	50 × 4.5[c]	LiChrosorb® (5)	E-1	2	UV (271)	7 = 5	—	—	1
B (2) U (1)	1	25 × 2.2	MicroPak® CH-10 (10)	E-2	0.5	UV (273)	4 = 4	Sulfadiazine (7)	—	2
B (2) U (1)	1	30 × 4	μ-Bondapak® C_{18} (10)	E-3	2	UV (271)	1 = 4	Cyclopenthiazide (5)	—	3

Thin-Layer Chromatography

Specimen (mℓ)	S	Plate (manufacturer)	Layer (mm)	Solvent	Post-separation treatment	Det. (nm)	R_f	Internal standard (R_f)	Other compounds (R_f)	Ref.
U (10)	3	NA (Laboratory)	Silica gel (0.25)	S-1	Sp: i. Conc. HCl; heat 100°C for 10 min. ii. Bratton-Marshall reagent[d]	Visual	2 = 0.98 4 = 0.26 5 = 0.94 7 = 0.52 8 = 0.78	—	—	4
U (10)	3	NA	Silica gel G (0.25)	S-2	Sp: Dimethylaminobenzaldehyde (25 g) in 80% acetone (75 mℓ) and ammonium hydroxide (25 mℓ)	Visual	2 = 0.82 6 = 0.61 7 = 0.19 8 = 0.46	—	—	5
D	3	20 × 20 cm (Laboratory)	Silica gel G (0.25)	S-3[e]	Sp: Dimethylaminobenzaldehyde (25 g) in 80% acetone (75 mℓ) and ammonium hydroxide (25 mℓ)	Visual	2 = 0.57 3 = 0.45 4 = 0.16 6 = 0.41 7 = 0.22	—	—	6

THIAZIDES[a] (continued)

1. Bemetizide; 2. Bendrofluazide; 3. Benzthiazide; 4. Chlorothiazide; 5. Cyclopenthiazide; 6. Cyclothiazide; 7. Hydrochlorothiazide; 8. Hydroflumethiazide

[a] See under hydrochlorothiazide also.
[b] See above for the identification of the thiazide corresponding to the number.
[c] Column temperature = 50°C.
[d] (i) 1% $NaNO_2$ in 1% H_2SO_4; (ii) 5% ammonium sulfamate; (iii) *N*-(1-naphthyl)ethylenediamine dihydrochloride (1% in 80% acetone).
[e] Alternative solvent systems and detection reagents are also described.

Note: E-1 = Ethanol + hexane.
30 : 70

E-2 = Methanol + 0.01 *M* acetic acid.
10 : 90

E-3 = Methanol + 0.01 *M* KH_2PO_4.
1 : 1

S-1 = Ethyl acetate + benzene.
8 : 2

S-2 = (i) Ethyl acetate + benzene; (ii) ethyl acetate + methanol + ammonia.
8 : 2 85 : 10 : 5

S-3 = Methyl ethyl ketone + *n*-hexane.
1 : 1

REFERENCES

1. Robinson, W. T. and Cosyns, L., *Clin. Biochem.*, 11, 172, 1978.
2. Shah, V. P., Prasad, V. K., Cabana, B. E., and Sojka, P., *Curr. Ther. Res.*, 24, 366, 1978.
3. Brodie, R. R., Chasseaud, L. F., Taylor, T., O'Kelly, D. A., and Darragh, A., *J. Chromatogr.*, 146, 152, 1978.
4. Osborne, B. G., *J. Chromatogr.*, 70, 190, 1972.
5. Sohn, D., Simon, J., Hanna, M. A., Ghali, G., and Tolba, R., *J. Chromatogr.*, 87, 570, 1973.
6. Stohs, S. J. and Scratchley, G. A., *J. Chromatogr.*, 114, 329, 1975.

THIAZINAMIUM METHYL SULFATE

Thin-Layer Chromatography

Specimen (mℓ)	S	Plate (manufacturer)	Layer (mm)	Solvent	Post-separation treatment	Det. (nm)	R_f	Internal standard (R_f)	Other compounds (R_f)	Ref.
U (1)	2	20 × 20 cm (Merck)	Silica gel 60 (0.25)	S-1[a]	D: Water (150) + sulfuric acid (150) + $FeCl_3$ (4 g) + acetone (800) + ethanol (800)	Transmission (525)	0.54	—	Thiazinamium sulfoxide (0.36)	1

Note: S-1 = i. Methanol

ii. Water + methanol + ammonia + ammonium acetate, pH 9.
42 : 200 : 6 : 8 g

[a] Bidimensional development. All developments and drying are carried out in the dark.

REFERENCE

1. Jonkman, J. H. G., Wijsbeek, J., Greving, J. E., Van Gorp, R. E., and de Zeeuw, R. A., *J. Chromatogr.*, 128, 208, 1976.

THIMEROSAL

High-Pressure Liquid Chromatography

Specimen (mℓ)	S	Column cm × mm	Packing (μm)	Elution	Flow rate (mℓ/min)	Det. (nm)	RT (min)	Internal standard (RT)	Other compounds (RT)	Ref.
D	3	50 × 3	Vydac® Anion-exchange resin (32—44)	E-1	6	UV (254)	3	—	—	1

Note: E-1 = 0.35% Perchloric acid in 0.001 *M* dibasic sodium phosphate, pH 7.

REFERENCE

1. **Fu, C.- C. and Sibley, M. J.,** *J. Pharm. Sci.*, **66**, 738, 1977.

THIOPENTAL

Gas Chromatography

Specimen (mℓ)	S	Column m × mm	Packing (mesh)	Oven temp (°C)	Gas (mℓ/min)	Det.	RT (min)	Internal standard (RT)	Deriv.	Other compounds (RT)	Ref.
B (5)	2	1.5×NA	3% PolyA-103 Gas Chrom® Q (80/100)	220	Nitrogen (45)	NPD	4.5	—	—	—	1
B (1)	2	1.8 × 2	3% SE-30 Chromosorb® W (80/100)	140	Nitrogen (40)	FID	5.8	Secobarbital (7)	Methyl[a]	—	2
B (1)	2	1.8 × 4	3% OV-17 Gas Chrom® Q (100/120)	195	Nitrogen (60)	ECD	3.6	Thiamylal (4.4)	Methyl[a]	—	3
B (2)	2	1.8 × 4	5% OV-1 Chromosorb® W (100/120)	205	Nitrogen (45)	FID	3.5	Butobarbitone (1.8)	—	—	4

High-Pressure Liquid Chromatography

Specimen (mℓ)	S	Column cm × mm	Packing (μm)	Elution	Flow rat (mℓ/min)	Det. (nm)	RT (min)	Internal standard (RT)	Other compounds (RT)	Ref.
B (0.5)	1	25 × 4.6	Partisil® 10/25 ODS (10)	E-1	0.5	UV (290)	8	Quinoline (10)	—	5
B (1)	1	30 × 3.9	μ-Bondapak® C_{18} (10)	E-2	2	UV (280)	3	—	—	6

THIOPENTAL (continued)

Thin-Layer Chromatography

Specimen (mℓ)	S	Plate (manufacturer)	Layer (mm)	Solvent	Post-separation treatment	Det. (nm)	R_f	Internal standard (R_f)	Other compounds (R_f)	Ref.
B (0.1)	2	20 × 20 cm (Merck)	Silica gel F_{254} (0.25)	S-1	—	UV Reflectance (287)	0.64	—	—	7

Note: E-1 = Methanol + 0.1% sodium acetate buffer, pH 6.5.
45 : 55

E-2 = Methanol + 0.02 *M* KCl, pH 2.
50 : 50

S-1 = Chloroform + acetone
19 : 1

[a] On column methylation with trimethylanilinium hydroxide.

REFERENCES

1. Sennello, L. T. and Kohn, F. E., *Anal. Chem.*, 46, 752, 1974.
2. Becker, K. E., *Anesthesiology*, 45, 656, 1976.
3. Smith, R. H., MacDonald, J. A., Thompson, D. S., and Flacke, W. E., *Clin. Chem.*, 23, 1306, 1977.
4. van Hamme, M. J. and Ghoneim, M. M., *Br. J. Anaesth.*, 50, 143, 1978.
5 Blackman, G. L., Jordan, G. J., and Paul, J. D., *J. Chromatogr.*, 145, 492, 1978.
6. Christensen, J. H. and Anderson, F., *Acta Pharmacol. Toxicol.*, 44, 260, 1979.
7. Gupta, R. N. and Eng, F., Unpublished data.

THIORIDAZINE

Gas Chromatography

Specimen (mℓ)	S	Column m × mm	Packing (mesh)	Oven temp (°C)	Gas (mℓ/min)	Det.	RT (min)	Internal standard (RT)	Deriv.	Other compounds (RT)	Ref.
B (4)	2	1.8 × 2	3% OV-17 Chromosorb® Q (100/120)	275	Helium (100)	FID	3.4	Chlorpromazine (0.6)	—	Mesoridazine (7.9) Thioridazine-3-sulfone (8.3) Thioridazine-*R*-sulfoxide (12.1)	1
B (5)	1	1 × 2	3% OV-17 Gas Chrom® Q (80/100)	264	Nitrogen (40)	FID	3.1	Prochlorperazine (2)	—	Mesordazine (7.1) Thioridazine-*S*-sulfone (7.5)	2

High-Pressure Liquid Chromatography

Specimen (mℓ)	S	Column cm × mm	Packing (μm)	Elution	Flow rate (mℓ/min)	Det. (nm)	RT (min)	Internal standard (RT)	Other compounds (RT)	Ref.
B (1)	2	25 × 2.8	Silica (9)	E-1	1.14	Fl[a] (Ex: 365, Fl: 440)	1.1	—	Northioridazine (1.5) Thioridazine-2-sulfone (2.5)	3
B (2)	2	25 × 4	μ-Bondapak® C_{18} (10)	E-2	2	UV (263)	8.5	Trifluoperazine (11.7)	Mesoridazine (3.9)	4

Thin-Layer Chromatography

Specimen (mℓ)	S	Plate (manufacturer)	Layer (mm)	Solvent	Post-separation treatment	Det. (nm)	R_f	Internal standard (R_f)	Other compounds (R_f)	Ref.
U (50)	3	NA	Silica gel (NA)	S-1	Sp: Perchloric acid	Visual	0.75	—	Mesoridazine (0.24) Northioridazine (0.8)	2

THIORIDAZINE (continued)

Note: E-1 = 2,2,4-Trimethylpentane + 2-aminopropane + acetonitrile + ethanol.
95.9 : 1 : 2.7 : 0.5

E-2 = Methanol + water + acetic acid + 1-heptane sulfonic acid.
66 : 34 : 1 : 0.0005 moles

S-2 = Chloroform + methanol + acetic acid.
9 : 1 : 1

[a] Post column oxidation with permanganate.

REFERENCES

1. Dinovo, E. C., Gottschalk, L. A., Nandi, B. R., and Geddes, P. G., *J. Pharm. Sci.*, 65, 667, 1976.
2. Ng, C. H. and Cranmer, J. L., *Br. J. Clin. Pharmacol.*, 4, 173, 1977.
3. Muusze, R. G. and Huber, J. F. K., *J. Chromatogr. Sci.*, 12, 779, 1974.
4. McCutcheon, J. R., *J. Anal. Toxicol.*, 3, 105, 1979.

THIOTHIXENE

High-Pressure Liquid Chromatography

Specimen (mℓ)	S	Column cm × mm	Packing (μm)	Elution	Flow rate (mℓ/min)	Det. (nm)	RT (min)	Internal standard (RT)	Other compounds (RT)	Ref.
D	3	1 × 2.1	Corasil® II (NA)	E-1	NA	UV (254)	1	*N*-(1-Naphthyl)-ethylenediamine (4.5)	—	1

Note: E-1 = Methanol + water + 95% ethanolamine.
280 : 40 : 0.037

REFERENCE

1. Wong, C. K., Cohen, D. M., and Munnelly, K. P., *J. Pharm. Sci.*, 65, 1090, 1976.

THYROXINE

High-Presure Liquid Chromatography

Specimen (mℓ)	S	Column cm × mm	Packing (μm)	Elution	Flow rate (mℓ/min)	Det. (nm)	RT (min)	Internal standard (RT)	Other compounds (RT)	Ref.
D	3	15 × 3.2	Nucleosil C_{18} (5)	E-1	0.7	Absorbance[a] (365)	10	—	Triiodothyronine (7.5)	1
D	3	30 × 4	μ-Bondapak® C_{18} (10)	E-2	2	UV (254)	30	—	Triiodothyronine (15) Reverse triiodothyronine (22)	2

Note: E-1 = Water + acetonitrile (+ 1% acetic acid).
5 : 2

E-2 = Methanol + 0.1% H_3PO_4.
50 : 50

[a] Post column redox reaction between cerium (IV) and arsenic (III), catalyzed by thyroxine.

REFERENCES

1. **Nachtmann, F., Knapp, G., and Spitzy, H.,** *J. Chromatogr.*, 149, 693, 1978.
2. **Hearn, M. T. W., Hancock, W. S., and Bishop, C. A.,** *J. Chromatogr.*, 157, 337, 1978.

TIAMENIDINE

Gas Chromatography

Specimen (mℓ)	S	Column m × mm	Packing (mesh)	Oven temp (°C)	Gas (mℓ/min)	Det.	RT (min)	Internal standard (RT)	Deriv.	Other compounds (RT)	Ref.
B (5)	1	2 × 0.8	5% OV-17 Glass beads (100/120)	Temp. Progr.	Helium (5)	MS-EI	3.3—4	[$^2H_4$]-Tiamenidine (3.3—4)	Heptafluorobutyryl	—	1
B (4)	1	1.5 × 2	3% OV-1 Chromosorb® W (100/120)	275	Helium (30)	MS-EI	2.8	Desmethyltiamenidine (2.6)	Benzyl	—	2

REFERENCES

1. Fehlhaber, H. -W., Metternich, K., Tripier, D., and Uihlein, M., *Biomed. Mass Spectrom.*, 5, 188, 1978.
2. Bryce, T. A. and Burrows, J. L., *Biomed. Mass Spectrom.*, 6, 27, 1979.

TICRYNAFEN

High-Pressure Liquid Chromatography

Specimen (mℓ)	S	Column cm × mm	Packing (μm)	Elution	Flow rate (mℓ/min)	Det. (nm)	RT (min)	Internal standard (RT)	Other compounds (RT)	Ref.
B (0.5)	1	25 × 2.6	ODS-HC-SIL-X-1 (NA)	E-1	0.75	UV (210)	6.3	Ethacrynic acid (9.7)	Dihydroticryna-fen (4.7)	1

Note: E-1 = Acetonitrile + 0.05 *M* phosphate buffer, pH 7.0.
22 : 78

REFERENCE

1. Randolph, W. C., Osborne, V. L., and Intoccia, A. P., *J. Pharm. Sci.,* 68, 1451, 1979.

TIENLIC ACID

Gas Chromatography

Specimen (mℓ)	S	Column m × mm	Packing (mesh)	Oven temp (°C)	Gas (mℓ/min)	Det.	RT (min)	Internal standard (RT)	Deriv.	Other compounds (RT)	Ref.
B (.05)	1	1.5 × 2	3% OV-225 Chromosorb® W (NA)	262	Nitrogen (45)	ECD	2.0	SKF-4260[a] A (3.0)	Methyl[b]	—	1

[a] 2-Chloro-10-(3′-dimethylaminopropyl) phenothiazine-*S*-oxide.
[b] With diazomethane.

REFERENCE

1. Desager, J. P., Vanderbist, M., Hwang, B., and Levandoski, P., *J. Chromatogr.*, 123, 379, 1976.

TIFLOREX

Gas Chromatography

Specimen (mℓ)	S	Column m × mm	Packing (mesh)	Oven temp (°C)	Gas (mℓ/min)	Det.	RT (min)	Internal standard (RT)	Deriv.	Other compounds (RT)	Ref.
B (0.5—2)	1	2 × 4	3% OV-17 Gas Chrom® Q (80/100)	210	Nitrogen (50)	ECD	8.6	Fenfluramine (4.6)	Trichlo-roacetyl	Nortiflorex (5.8)	1

REFERENCE

1. Bianchetti, G., Mitchard, M., and Morselli, P. L., *J. Chromatogr.*, 152, 87, 1978.

TILIDINE

Gas Chromatography

Specimen (mℓ)	S	Column m × mm	Packing (mesh)	Oven temp (°C)	Gas (mℓ/min)	Det.	RT (min)	Internal standard (RT)	Deriv.	Other compounds (RT)	Ref.
B, U (1—5)	1	1.8 × 2	1% CRS 101 + 1.5% HI-EFF-1BP Gas Chrom® Q (100/120)	165	Nitrogen (20)	NPD	3.7	a (11)	—	Nortilidine (4.8) Bisnortilidine (6.7)	1

[a] 1[(R)-*N*-(dipropyl)-amino]-4-phenyl-4-ethoxycarbonylcyclohexene.

REFERENCE

1. Hengy, H., Vollmer, K.- O., and Gladigau, V., *J. Pharm. Sci.*, 67, 1765, 1978; *Clin. Chem.*, 24, 692, 1978.

TIMOLOL

Gas Chromatography

Specimen (mℓ)	S	Column m × mm	Packing (mesh)	Oven temp (°C)	Gas (mℓ/min)	Det.	RT (min)	Internal standard (RT)	Deriv.	Other compounds (RT)	Ref.
B (1) U (0.1)	1	1.8 × NA	1% OV-17 Gas Chrom® Q (80/100)	185	Helium (75)	ECD	7.6	Desmethyltimo- lol (6)	Heptafluo- robutyryl	—	1

REFERENCE

1. Tocco, D. J., de Luna, F. A., and Duncan, A. E. W., *J. Pharm. Sci.*, 64, 1879, 1975.

TINIDAZOLE

Gas Chromatography

Specimen (mℓ)	S	Column m × mm	Packing (mesh)	Oven temp (°C)	Gas (mℓ/min)	Det.	RT (min)	Internal standard (RT)	Deriv.	Other compounds (RT)	Ref.
B (NA) Tissue	2	1 × 2	3% OV-11 Gas Chrom® Q (80/100)	215	Nitrogen (20)	NPD	3.8	4-Nitrodiphenylamine (5.1)	—	—	1

High-Pressure Liquid Chromatography

Specimen (mℓ)	S	Column cm × mm	Packing (μm)	Elution	Flow rate (mℓ/min)	Det. (nm)	RT (min)	Internal standard (RT)	Other compounds (RT)	Ref.
B (0.3)	2	100 × 2.2	ETH-Permaphase (25—37)	E-1	1	UV (315)	3	—	—	2
B (1)	2	20 × 3.9	μ-Bondapak®-phenyl (10)	E-2	2	UV (313)	9	—	Hydroxy metabolite[a] (6.1)	3

Note: E-1 = Hexane + chloroform + ethanol.
90 : 15 : 0.5

E-2 = 0.05 *M* Phosphate buffer, pH 7 + methanol.
86 : 14

[a] Ethyl [2-(2-hydroxymethyl-5-nitro-1-imidazolyl)-ethyl] sulfone.

REFERENCES

1. **Laufen, H., Scharpf, F., and Bartsch, G.,** *J. Chromatogr.*, 163, 217, 1979.
2. **Nachbaur, J. and Joly, H.,** *J. Chromatogr.*, 145, 325, 1978.
3. **Alton, K. B. and Patrick, J. E.,** *J. Pharm. Sci.*, 68, 599, 1979.

TIZOLEMIDE

High-Pressure Liquid Chromatography

Specimen (mℓ)	S	Column cm × mm	Packing (μm)	Elution	Flow rate (mℓ/min)	Det. (nm)	RT (min)	Internal standard (RT)	Other compounds (RT)	Ref.
B (NA)	1	12.5 × 4	Nucleosil 7-C_{18} (7)	E-1	2	UV (202)	1.8	Isopropylamino analog	M_2[a] (2.6) M_3[b] (1.4)	1, 2

Thin-Layer Chromatography

Specimen (mℓ)	S	Plate (manufacturer)	Layer (mm)	Solvent	Post-separation treatment	Det. (nm)	R_f	Internal standard (R_f)	Other compounds (R_f)	Ref.
U (1)	2	NA[c] (Merck)	Silica gel F_{254} (NA)	S-1	—	UV Reflectance (228)	0.6	—	M_2[a] (0.4) M_3[b] (0.3)	2

Note: E-1 = 0.067 *M* Phosphate buffer + methanol.
1 : 1

S-1 = Chloroform + cyclohexane + acetic acid + methanol.
7.5 : 7.5 : 0.1 : 85

[a] 2-Chloro-5-[2,3,dihydro-3-methyl-2-(methylimino)-4-thiazolyl] benzene sulfonamide.
[b] 2-Chloro-5[2,3,dihydro-3-methyl-2-imino-4-thiazolyl] benzene sulfonamide.
[c] Merck Cat. #5715.

REFERENCES

1. Uihlein, M., *Chromatographia,* 12, 408, 1979.
2. Sistovaris, N. and Uihlein, M., *J. Chromatogr.,* 167, 109, 1978.

TOCAINIDE

High-Pressure Liquid Chromatography

Specimen (mℓ)	S	Column cm × mm	Packing (μm)	Elution	Flow rate (mℓ/min)	Det. (nm)	RT (min)	Internal standard (RT)	Other compounds (RT)	Ref.
B (1)	2	15 × 4.5	Partisil® (5)	E-1	1	UV (230)	5.8	—	—	1

Note: E-1 = Dichloromethane + methanol + 1 *M* perchloric acid.
89.5 : 10 : 0.5

[a] Different conditions for the analysis of a number of antiarrhythmic drugs are described.

REFERENCE

1. **Lagerstrom, P.- O. and Persson, B.- A.,** *J. Chromatogr.*, 149, 331, 1978.

TOLAZAMIDE

Gas Chromatography

Specimen (mℓ)	S	Column m × mm	Packing (mesh)	Oven temp (°C)	Gas (mℓ/min)	Det.	RT (min)	Internal standard (RT)	Deriv.	Other compounds (RT)	Ref.
B (0.1—0.5)	2	2 × 3	5% SE-30 Chromosorb® W (80/100)	Temp. Progr.	Helium (40)	ECD	8.8	Tolbutamide (5.2)	i. Methyl[a] ii. Trifluoroacetyl	Carbutamide (7.2) Glycodiazin (10.4)	1

[a] Wtih diazomethane.

REFERENCE

1. Schlicht. H.- J., Gelbke, H.-P., and Schmidt, G., *J. Chromatogr.*, 155, 178, 1978.

TOLBUTAMIDE

Gas Chromatography

Specimen (mℓ)	S	Column m × mm	Packing (mesh)	Oven temp (°C)	Gas (mℓ/min)	Det.	RT (min)	Internal standard (RT)	Deriv.	Other compounds (RT)	Ref.
B (0.5)	1	1.8 × 2	3% OV-17 Chromosorb® W (100/120)	210	Methane 5 + Argon 95 (30)	ECD	—	i. Chlorpropamide (5) ii. T-isoprop[a] (4)	i. Hydrolysis[b] ii. 2,4-Dinitrophenyl	T-OH[c] (6) T-COOH[d] (6)	1[e]
B (1)	1	1.5 × 2	1% SP2100 Supelcoport (100/120)	170	Helium (30)	MS-EI[f]	6.2	Chlorpropamide (4.7)	i. Methyl ii. Trifluoroacetyl	T-OH (7)[h] T-COOH (8.5)[h]	3

High-Pressure Liquid Chromatography

Specimen (mℓ)	S	Column cm × mm	Packing (μm)	Elution	Flow rate (mℓ/min)	Det. (nm)	RT (min)	Internal standard (RT)	Other compounds (RT)	Ref.
B (1)	1	100 × NA	ETH-Permaphase (NA)	E-1	0.4	UV (254)	6	*N*-(*p*-Methoxybenzenesulfonyl)-*N*′-cyclohexylurea (9)	—	4
B (0.2)	2	30 × 4	μ-Bondapak® C_{18} (10)	E-2	2.2	UV (254)	5	1-Isopentyl-3-*p*-tolylsulfonylurea (8)	Chlorpropamide (3)	5
B (0.1)	1	30 × 4	μ-Bondapak® C_{18} (10)	E-3	1.5	UV (200)	5.7	—	T-COOH[d] (3.2)	6

TOLBUTAMIDE (continued)

Thin-Layer Chromatography

Specimen (mℓ)	S	Plate (manufacturer)	Layer (mm)	Solvent	Post-separation treatment	Det. (nm)	R_f	Internal standard (R_f)	Other compounds (R_f)	Ref.
B (0.25)	2	20 × 20 cm (Meck)	Silica gel 60 F_{254} (0.25)	S-1	—	UV Reflectance (228)	0.66	—	Chlorpropamide (0.25)	7

Note: E-1 = Methanol + 0.01 *M* monobasic sodium citrate.
30 : 70

E-2 = Acetonitrile + 1% acetic acid, pH 5.5
28 : 72

E-3 = Acetonitrile + 0.05% phosphoric acid.
45 : 55

S-1 = Chloroform + methanol.
19 : 1

[a] 1-Isopropyl-3-*p*-carboxyphenylsulfonylurea.
[b] Chlorpropamide; T-isoprop; T-OH and T-COOH produce, on hydrolysis and reaction with 2,4-dinitrofluorobenzene, propyl- ; isopropyl- ; butyl- ; and butyl-2,4-dinitroanilines, respectively.
[c] 1-Butyl-3-*p*-hydroxymethylphenylsulfonylurea.
[d] 1-Butyl-3-*p*-carboxyphenylsulfonylurea.
[e] The senior author has also described a chemical ionization procedure without prior separation by gas chromatography (Reference 2).
[f] An FID was also used.
[g] With diazomethane.
[h] Under different conditions.

REFERENCES

1. **Matin, S. B. and Rowland, M.,** *Anal. Lett.*, 6, 865, 1973.

2. Knight, J. B. and Matin, S. B., *Anal. Lett.*, 7, 529, 1974.
3. Braselton, W. E., Jr., Ashline, H. C., and Bransome, E. D., Jr., *Anal. Lett.*, 8, 301, 1975.
4. Weber, D. J., *J. Pharm. Sci.*, 65, 1502, 1976.
5. Hill, R. E. and Crechiolo, J., *J. Chromatogr.*, 145, 165, 1978.
6. Nation, R. L., Peng, G. W., and Chiou, W. L., *J. Chromatogr.*, 146, 121, 1978.
7. Gupta, R. N. and Eng, F., Unpublished data.

TOLMETIN

Gas Chromatography

Specimen (mℓ)	S	Column m × mm	Packing (mesh)	Oven temp (°C)	Gas (mℓ/min)	Det.	RT (min)	Internal standard (RT)	Deriv.	Other compounds (RT)	Ref.
B, U (1)	1	1.5 × NA	3% OV-17 Gas Chrom® Q (100/120)	230	Nitrogen (40)	FID	5	a (6.25)	Methyl[b]	c (13.5)	1
B (1)	1	1.8 × 4	3% OV-1 Gas Chrom® Q (60/80)	200	Nitrogen (60)	FID	6	a (8)	Methyl[b]	—	2
B (0.1)	1	0.6 × 4	3% XE-60 Gas Chrom® Q (80/100)	230	Methane 5 + Argon 95 (60)	ECD	3.8	Dichloro analog (6.7)	Pentafluorobenzyl	—	3

High-Pressure Liquid Chromatography

Specimen (mℓ)	S	Column cm × mm	Packing (μm)	Elution	Flow rate (mℓ/min)	Det. (nm)	RT (min)	Internal standard (RT)	Other compounds (RT)	Ref.
D	3	25 × 2.2	LiChrosorb® SI-60 (10)	E-1 Gradient[e]	2	UV (254)	8.6	—	d	4

Note: E-1 = i. Acetic acid + hexane.
0.25 : 100
ii. Isopropanol + hexane + acetic acid
50 : 50 : 25

[a] 1,4-Dimethyl-5-p-chlorobenzoyl-pyrrole-2-propionic acid.
[b] With diazomethane.
[c] 1-Methyl-5-*p*-carboxybenzoyl-pyrrole-2-acetic acid.

[d] Retention times of different intermediates which could be present as impurities are given.
[e] Concave gradient no. 7.

REFERENCES

1. Selley, M. L., Thomas, J., and Triggs, E. J., *J. Chromatogr.*, 94, 143, 1974.
2. Cressman, W. A., Lopez, B., and Sumner, D., *J. Pharm. Sci.*, 64, 1965, 1975.
3. Ng, K.-T., *J. Chromatogr.*, 166, 527, 1978.
4. Gilpin, R. K. and Janicki, C. A., *J. Chromatogr.*, 147, 501, 1978.

TRANYLCYPROMINE

Gas Chromatography

Specimen (mℓ)	S	Column m × mm	Packing (mesh)	Oven temp (°C)	Gas (mℓ/min)	Det.	RT (min)	Internal standard (RT)	Deriv.	Other compounds (RT)	Ref.
B (2)	1	1.8 × 2	2% OV-1 Chromosorb® G (100/120)	185	Methane 5 + Argon 95 (33)	ECD	3.1	*N*-*n*-Propylamphetamine (4.1)	Trichloroacetyl	—	1

Thin-Layer Chromatography

Specimen (mℓ)	S	Plate (manufacturer)	Layer (mm)	Solvent	Post-separation treatment	Det. (nm)	R_f	Internal standard (R_f)	Other compounds (R_f)	Ref.
B (5)	2	20 × 20 cm (Merck)	Silica gel 60 (NA)	S-1	Sp: Triethanolamine-isopropanol (20/:80, V/V)	Fl[a] (Ex: 365, Fl: 546)	0.59	—	—	2

Note: S-1 = Cyclohexane + isoamyl alcohol.
85 : 15

[a] Dansyl derivatives are prepared prior to chromatography.

REFERENCES

1. Baselt, R. C., Stewart, C. B. and Shaskan, E., *J. Anal. Toxicol.*, 1, 215, 1977.
2. Lang, A., Geisler, H. E., and Mutschler, E., *Arzneim. Forsch.*, 28(1), 575, 1978.

TRH

(Thyrotropin-Releasing Hormone)

High-Pressure Liquid Chromatography

Specimen (mℓ)	S	Column cm × mm	Packing (μm)	Elution	Flow rate (mℓ/min)	Det. (nm)	RT (min)	Internal standard (RT)	Other compounds (RT)	Ref.
D	3	30 × 3.9	μ-Bondapak®[a] C_{18} (10)	E-1	2	UV (210)	12	—	Pyroglutamylhistidyl-proline (9) Pyroglutamyl-3-methyl-histidyl-prolineamide (15) *L-N*-(2-oxopiperidin-6-ylcarbonyl)-1-histidyl-1-thiazlidine-4-carboxamide (19) Pyroglutamyl-phenylalanyl-prolineamide (27) Glutamylhistidyl-prolineamide (62)	1

Note: E-1 = Acetonitrile + 0.1% heptanesulfonic acid in 0.02 *N* acetic acid.
6.5 : 93.5

[a] Column temperature = 60°C.

TRH (continued)

(Thyrotropin-Releasing Hormone)

REFERENCE

1. Spindel, E. and Wurtman, R. J., *J. Chromatogr.*, 175, 198, 1979.

TRIAMTERENE

High-Pressure Liquid Chromatography

Specimen (mℓ)	S	Column cm × mm	Packing (μm)	Elution	Flow rate (mℓ/min)	Det. (nm)	RT (min)	Internal standard (RT)	Other compounds (RT)	Ref
B, U (0.5)	1	25 × 3.2	LiChrosorb SI-60 (5)	E-1	2	Fl (Ex: 335, Fl: 470)	3.8	—	Metabolite (5)	1

Thin-Layer Chromatography

Specimen (mℓ)	S	Plate (manufacturer)	Layer (mm)	Solvent	Post-separation treatment	Det. (nm)	R_f	Internal standard (R_f)	Other compounds (R_f)	Ref.
B (1) U (0.01)	2	NA (Merck)	Silica gel G-60 (NA)	S-1	—	Fl (Ex: 365, Fl: 440)	0.65	—	Hydroxytriamterene (0.5)	2

Note: E-1 = Dichloromethane + hexane + methanol + 70% perchloric acid.
57 : 35 : 8 : 0.1

S-1 = Ethyl acetate + methanol + ammonia.
60 : 30 : 10

TRIAMITERENE (continued)

REFERENCES

1. Sved, S., Sertie, J. A. A., and McGilveray, I. J., *J. Chromatogr.*, 162, 474, 1979.
2. Grebian, B., Geisler, H. E., and Mutschler, E., *Arzneim. Forsch.*, 26, 2125, 1976.

TRIBROMSALAN

Thin-Layer Chromatography

Specimen (mℓ)	S	Plate (manufacturer)	Layer (mm)	Solvent	Post-separation treatment	Det. (nm)	R_f	Internal standard (R_f)	Other compounds (R_f)	Ref.
B (5)	2	20 × 20 cm (Laboratory)	Diethylamino-ethyl cellulose (0.1)	S-1	—	Fl (Ex: 350, Fl: 390)	0.43	—	—	1

Note: S-1 = Methanol + acetic acid.
92.5 : 7.5

REFERENCE

1. **Hong, H. S. C., Steltenkamp, R. J., and Smith, N. L.,** *J. Pharm. Sci.,* 64, 2007, 1975.

TRICLOCARBAN

Gas Chromatography

Specimen (mℓ)	S	Column m × mm	Packing (mesh)	Oven temp (°C)	Gas (mℓ/min)	Det.	RT (min)	Internal standard (RT)	Deriv.	Other compounds (RT)	Ref.
B (2)	2	1.5 × 4	1.5% SP-2250 + 1.95% SP-2401 Supelcon (100/120)	210	Nitrogen (75)	ECD	7.8	—	Trimethyl-silyl	—	1

REFERENCE

1. Hoar, D. R. and Bowen, M. H., *J. Pharm. Sci.*, 66, 725, 1977.

TRIFLUBAZAM

High-Pressure Liquid Chromatography

Specimen (mℓ)	S	Column cm × mm	Packing (μm)	Elution	Flow rate (mℓ/min)	Det. (nm)	RT (min)	Internal standard (RT)	Other compounds (RT)	Ref.
B (1) U(5)	1	1 × 2.1[a]	Pelosil® HC (NA)	E-1	3	UV (254)	1	11-α-Hydroxy-17-α-methyl-testosterone (4.5)	Nortiflubazam (2.5) Hydroxyfluba-zam (6.5) Hydroxynorflu-bazam (13)	1

Note: E-1 = Dioxane + isoctane.
15 : 85

[a] Column temperature = 30°C.

REFERENCE

1. **Huettemann, R. E. and Shroff, A. P.,** *J. Pharm. Sci.*, 64, 1339, 1975.

TRIFLUOROACETYLADRIAMYCIN

Thin-Layer Chromatography

Specimen (mℓ)	S	Plate (manufacturer)	Layer (mm)	Solvent	Post-separation treatment	Det. (nm)	R_f	Internal standard (R_f)	Other compounds (R_f)	Ref.
B (0.3)	2	20 × 20 cm (Merck)	Silica gel (0.25)	S-1	—	Fl (Ex: 475, Fl: 580)	0.43	—	Daunomycinone (0.79) Adriamycinone (0.62) *N*-Trifluoroacetyl adriamycinol (0.14)	1

Note: S-1 = Chloroform + methanol + acetic acid.
93 : 5 : 2

REFERENCE

1. Barbieri, B., Abruzzi, R., Benigni, A., Rizzadini, M., Donelli, M. G., Garattini, S., and Salmona, M., *J. Chromatogr.*, 163, 195, 1979.

TRIHEXYPHENIDYL

Gas Chromatography

Specimen (mℓ)	S	Column m × mm	Packing (mesh	Oven temp (°C)	Gas (mℓ/min)	Det.	RT (min)	Internal standard (RT)	Deriv.	Other compounds (RT)	Ref.
D	3	1.8 × 4	3% OV-1 Chromosorb® W (100/120)	210	Nitrogen (60)	FID	6.8	*n*-Tricosane (8.8)	—	Tridihexethyl (4)	1
D	3	1.8 × 4	3% OV-210 Chromosorb® W (100/120)	170	Nitrogen (38)	FID	8.4	—	—	a	2

Thin-Layer Chromatography

Specimen (mℓ)	S	Plate (manufacturer)	Layer (mm)	Solvent	Post-separation treatment	Det. (nm)	R_f	Internal standard (R_f)	Other compounds (R_f)	Ref.
D	3	20 × 20 cm (Merck)	Silica gel F_{254} (0.25)	S-1	Sp: Dragendorff reagent	Visual	0.46	—	a	2

Note: S-1 = Chloroform + ethylacetate + methanol + acetone.
20 : 20 : 20 : 40

[a] Impurities from the manufacturing process are identified in different commercial preparations.

REFERENCES

1. Bargo, E., *J. Pharm. Sci.*, 68, 503, 1979.
2. Poirier, M. A., Curran, N. M., McErlane, K. M., and Lovering, E. G., *J. Pharm. Sci.*, 68, 1124, 1979.

TRIMETHADIONE

Gas Chromatography

Speimen (mℓ)	S	Column m × mm	Packing (mesh)	Oven temp (°C)	Gas (mℓ/min)	Det.	RT (min)	Internal standard (RT)	Deriv.	Other compounds (RT)	Ref.
B (2)	2	1.2 × 4	3% OV-17 Gas Chrom® Q (100/120)	100	Helium (60)	FID	1.6	Paramethadione (2.5)	—	Dimethadione (4.7)	1

REFERENCE

1. Booker, H. E. and Darcey, B., *Clin. Chem.*, 17, 607, 1971.

TRIMETHOPRIM

Gas Chromatography

Specimen (mℓ)	S	Column m × mm	Packing (mesh)	Oven temp (°C)	Gas (mℓ/min)	Det.	RT (min)	Internal standard (RT)	Deriv.	Other compounds (RT)	Ref.
B (0.5)	1	1.8 × 4	10% Poly S-179 Chromosorb® W (80/100)	330	Helium (45)	NPD	8.5	BW-214U[a] (6.3)	—	—	1

High-Pressure Liquid Chromatography

Specimen (mℓ)	S	Column cm × mm	Packing (μm)	Elution	Flow rate (mℓ/min)	Det. (nm)	RT (min)	Internal standard (RT)	Other compounds (RT)	Ref.
B (0.02) U (0.01)	1	15 × 4.6	LiChrosorb® RP8 (5)	E-1	1.2	UV (225)	5.5	—	Sulfamethoxazole (6.3) N^4-Acetylsulfamethoxazole (7.5)	2
B (1)	1	25 × 4	LiChrosorb® RP8 (10)	E-2	1	Fl (Ex: 279, Fl: 370)	12	—	—	3
B (1)	1	25 × 4.6	Spherisorb® ODS (10)	E-3	1	UV (254)	6	—	—	4
B (2) U (0.1)	1	30 × 3.9	μ-Porasil® (10)	E-4	1.5	UV (280)	5.3	b (4.8)	Trimethoprim-*N*-1-oxide (4.8) α-Hydroxytrimethoprim (18) Trimethoprim-*N*-3-oxide (29)	5

TRIMETHOPRIM (continued)

Note: E-1 = Methanol + 0.067 *M* KH_2PO_4 + 0.067 *M* Na_2HPO_4 pH 6.7.
80 : 390 : 10

E-2 = Methanol + 0.015 *M* sodium borate pH 9.
35 : 65

E-3 = Acetonitrile + 0.1 *M* KH_2PO_4 pH 2.5 + acetic acid + ethyl acetate.
30 : 70 : 1 : 1

[a] 2,4-Diamino-5-(3′, 4′-dibromophenyl)-6-methylpyrimidine.
[b] 2,4-Diamino-5-(3,5-dimethoxy-4-methylbenzyl)-pyrimidine.

REFERENCES

1. Land, G., Dean, K., and Bye, A., *J. Chromatogr.*, 146, 143, 1978.
2. Vree, T. B., Hekster, Y. A., Baars, A. M., Damsma, J. E., and van der Kleijn, E., *J. Chromatogr.*, 146, 103, 1978.
3. Gautam, S. R., Chungi, V. S., Bourne, D. W. A., and Munson, J. W., *Anal. Lett.*, B11, 967, 1978.
4. Bury, R. W. and Mashford, M. L., *J. Chromatogr.*, 163, 114, 1979.
5. Weinfeld, R. E. and Macasieb, T. C., *J. Chromatogr.*, 164, 73, 1979.

N-N'-TRIMETHYLENE-BIS-(PYRIDINIUM-4-ALDOXIME)

High-Pressure Liquid Chromatography

Specimen (mℓ)	S	Column cm × mm	Packing (μm)	Elution	Flow rate (mℓ/min)	Det. (nm)	RT (min)	Internal standard (RT)	Other compounds (RT)	Ref.
D	3	30 × 3.9	μ-Bondapak® C_{18} (10)	E-1	1.5	UV (254)	2.5	—	Atropine (3) Methylparaben (4) Benactyzine (6.8) Propylparaben (8.2)	1

Note: E-1 = Acetonitrile + 4% PIC B-7 reagent in deionized water.
35 : 65

REFERENCE

1. **Brown, N. D., Hall, L. N., Sleeman, H. K., Doctor, B.P., and Demaree, G. E.,** *J. Chromatogr.,* 148, 453, 1978.

TRIMIPRAMINE

Gas Chromatography

Specimen (mℓ)	S	Column m × mm	Packing (mesh)	Oven temp (°C)	Gas (mℓ/min)	Det.	RT (min)	Internal standard (RT)	Deriv.	Other compounds (RT)	Ref.
B (2)	2	1.8 × 4	3% OV-17 Gas Chrom® Q (100/120)	240	Nitrogen (30)	NPD	2.8	N-7084[a] (6.5) Benzoctamine (10.8)	Acetyl	Desmethyltrimipramine (15.1)	1

High-Pressure Liquid Chromatography

Specimen (mℓ)	S	Column cm × mm	Packing (μm)	Elution	Flow rate (mℓ/min)	Det. (nm)	RT (min)	Internal standard (RT)	Other compounds (RT)	Ref.
B (1)	2	15 × 3	Partisil® 5 (5)	E-1	1.0	UV (254)	3.1	Promazine (5.5)	b	2

Thin-Layer Chromatography

Specimen (mℓ)	S	Plate (manufacturer)	Layer (mm)	Solvent	Post-separation treatment	Det. (nm)	R_f	Internal standard (R_f)	Other compounds (R_f)	Ref.
B (2)	2	20 × 20 cm (Merck)	Silica gel (0.25)	S-1	Ex: i. HCl vapors ii. HNO_2 vapors	Visible reflectance (430)	0.89	Imipramine (0.59)	Desmethyltrimipramine (0.40)	1

Note: E-1 = Dichloromethane + methanol + buffer, pH 3.2.
90 : 10 : 0.15

S-1 = Ethyl acetate + methanol + ammonia.
85 : 8 : 2

[a] Dimethyl group of amitriptyline replaced by pyrrolidine.
[b] Analysis of a number of other tricyclic antidepresants is described.

REFERENCES

1. Gupta, R. N. and Gupta, M. L., unpublished data.
2. Uges, D. R. A. and Bouma, P., *Pharm. Weekbl.*, 1, 417, 1979.

TRIPROLIDINE

Thin-Layer Chromatography

Specimen (mℓ)	S	Plate (manufacturer)	Layer (mm)	Solvent	Post-separation treatment	Det. (nm)	R_f	Internal standard (R_f)	Other compounds (R_f)	Ref.
B (1)	1	20 × 20 cm (Merck)	Silica gel (0.25)	S-1	Sp: 2 *M* Aqueous ammonium bisulfate	Fl reflectance (Ex: 300, Fl: 405)	0.62	—	—	1

Note: S-1 = Methanol + ammonium hydroxide + chloroform.
10 : 1 : 89

REFERENCE

1. DeAngelis, R. L., Kearney, M. F., and Welch, R. M., *J. Pharm Sci.*, 66, 841, 1977.

TRIS (HYDROXYMETHYL) AMINOMETHANE

Gas Chromatography

Specimen (mℓ)	S	Column m × mm	Packing (mesh)	Oven temp (°C)	Gas (mℓ/min)	Det.	RT (min)	Internal standard (RT)	Deriv.	Other compounds (RT)	Ref.
B (0.1)	2	1 × NA	1% OV-17 Gas Chrom® Q (100/120)	255	Nitrogen (40)	FID	5	1,2,6-Hexane-triol (10)	Benzoyl	—	1

REFERENCES

1. **Hulshoff, A. and Kostenbauder, H. B.,** *J. Chromatogr.*, 145, 155, 1978.

VALPROIC ACID

Gas Chromatography

Specimen (mℓ)	S	Column m × mm	Packing (mesh)	Oven temp (°C)	Gas (mℓ/min)	Det.	RT (min)	Internal standard (RT)	Deriv.	Other compounds (RT)	Ref.
B (0.05)	2	1.5 × 3	10% HI-EFF-1AP + 2% phosphoric acid Diatomite (80/100)	150	Nitrogen (30)	FID	4.2	Octanoic acid (7.5)	—	—	1
B (1)	2	1.8 × 4	10% Carbowax® 20 *M*-terphthalic acid Chromosorb® W (80/100)	215	Nitrogen (45)	FID	3.5	Methylmyristate (4.5)	—	—	2
B (1)	2	1.8 × 2	(NA) FFAP Gas Chrom® Q (80/100)	170	Nitrogen (30)	FID	2.8	Caprylic acid (4.3)	—	—	3
B (0.02)	2	1.8 × NA	5% DEGS-PS Supelcoport (100/120)	135	Nitrogen (30)	FID	2	2-Ethyl-2-methylcaproic acid (2.5)	—	—	4
B, U (1)	1	1.5 × 6	3% OV-17 Gas Chrom® Q (80/100)	93	Nitrogen (45)	FID	2.7	Cyclohexane carboxylic acid (4.2)	Methyl	—	5
B (0.1)	2	1.5 × 4	1.5% SP-1000 Gas Chrom® Q (60/80)	140	Nitrogen (60)	FID	4.4	Caproic acid (NA)	—	—	5
B (0.5)	2	1.5 × 3	10% AT-1000 Gas Chrom® Q (100/120)	200	Nitrogen (22)	FID	5.5	2-Ethyl-2-methylcaproic acid (6)	—	—	6
B (0.1)	2	1.8 × 2	2% SP-1000 Supelcoport (100/120)	80	Nitrogen (20)	FID	1.5	Heptanoic acid (2)	Methyl[b]	—	7
B (0.5)	2	1.8 × 2	5% OV-17[c]	Temp.	Nitrogen	FID	3.8	Cyclohexyla-	Methyl[d]	—	8

			Gas Chrom® Q (100/120)	Progr.	(40)			cetic acid			
B (1)	2	1 × 2	3% OV-225 Gas Chrom® Q (100/120)	110	Helium (30)	FID	1.2	Paramethadione (1.6)	—	—	9
B (0.2)	2	1.5 × 2	5% FFAP Gas Chrom® W (80/100)	140	Helium (30)	FID	4	4-Methylvaleric acid (2.5)	—	—	10
B (0.05)	2	1.5 × 2	3% OV-17 Chromosorb® W (100/120)	115	Nitrogen (20)	FID	2.8	Octanoic acid (53)	Butyl[e]	—	11
B (0.1)	2	2 × 2 (Steel)	5% Carbowax®-20 *M* Chromosorb® P (80/100)	150	Nitrogen[f] (45)	FID	7	Octanoic acid (12)	—	—	12
B (0.25)	2	2 × 2	3% OV-17 Gas Chrom® Q (80/100)	205	Nitrogen (30)	FID	2.9	Cyclohexane carboxylic acid (4.8)	Phenacyl	—	13[g]
B (1)	2	1.5 × 2	10% SP-1000 Chromosorb® W (100/120)	160	Nitrogen (50)	FID	3.5	Caproic acid (2.2)	—	—	14

High-Pressure Liquid Chromatography

Specimen (mℓ)	S	Column cm × mm	Packing (μm)	Elution	Flow rate (mℓ/min)	Det. (nm)	RT (min)	Internal standard (RT)	Other compounds (RT)	Ref.
B (0.5)	2	30 × 4	μ-Bondapak® C_{18} (10)	E-1	0.7	Absorbance[h] (425)	7.2	5-(*p*-Methylphenyl)-5-phenylhydantoin (11.2)	—	15
B (0.25)	2	25 × 4.6	LiChrosorb® RP-18 (10)	E-2	2.5	UV[i] (246)	4.5	Cyclohexane carboxylic acid (2.8)	—	16

VALPROIC ACID (continued)

Note: E-1 = Methanol + water.
64 : 36

E-2 = Methanol + water.
75 : 25

[a] On-column methylation with trimethylanilinium hydroxide.
[b] Extractive alkylation in the presence of iodomethane and tetrabutyl ammonium hydrogen sulfate.
[c] The column is primed by injecting 6*N* HCl in ethyl acetate.
[d] With methanol and HCl.
[e] With butyl iodide in the presence of tetramethylammonium hydroxide and dimethyl acetamide.
[f] Saturated with formic acid.
[g] Nitrophenacyl esters have been analyzd with a nitrogen detector.
[h] After post column reaction with bromocresol purple indicator.
[i] Phenacyl esters are prepared prior to chromatography.

REFERENCES

1. **Fellenberg, A. J. and Pollard, A. C.,** *Clin. Chim. Acta,* 81, 203, 1977.
2. **Dusci, L. J. and Hackett, L. P.,** *J Chromatogr.,* 132, 145, 1977.
3. **Jensen, C. J. and Gugler, R.,** *J. Chromatogr.,* 137, 188, 1977.
4. **Jacobs, C., Bojasch, M., and Hanefeld, F.,** *J. Chromatogr.,* 146, 494, 1978.
5. **Berry, D. J. and Clarke, L. A.,** *J. Chromatogr.,* 156, 301, 1978.
6. **Libeer, J.-C., Scharpi, S., Schepens, P., and Verkerk, R.,** *J. Chromatogr.,* 160, 285, 1978.
7. **Gyllenhall, O. and Albinsson, A.,** *J. Chromatogr.,* 161, 343, 1978.
8. **Tupper, N. L., Solow, E. B., and Kenfield, C. P.,** *J. Anal. Toxicol.,* 2, 203, 1978.
9. **Balkon, J.,** *J. Anal. Toxicol.,* 2, 207, 1978; 3, 78, 1979.
10. **Levy, R. H., Martis, L., and Lai, A. A.,** *Anal. Lett.,* B11, 257, 1978.
11. **Hulshoff, A. and Roseboom, H.,** *Clin. Chim. Acta,* 93, 9, 1979.
12. **Pilerie, B.,** *J. Chromatogr.,* 162, 446, 1979.

13. Gupta, R. N., Eng, F., and Gupta, M. L., *Clin. Chem.*, 25, 1303, 1979.
14. Peyton, G. A., Harris, S. C., and Wallace, J. E., *J. Anal. Toxicol.*, 3, 108, 1979.
15. Farinotti, R. and Mahuzier, G., *J. Liq. Chromatogr.*, 2, 345, 1979.
16. Gupta, R. N., Keane, P. M., and Gupta, M. L., *Clin. Chem.*, 25, 1984, 1979.

VANCOMYCIN

High-Pressure Liquid Chromatography

Specimen (mℓ)	S	Column cm × mm	Packing (μm)	Elution	Flow rate (mℓ/min)	Det. (nm)	RT (min)	Internal standard (RT)	Other compounds (RT)	Ref.
D	3	30 × 4	μ-Bondapak® C_{18} (10)	E-1	1	UV (225)	6	—	Anisomycin (14) Trimethoprim (17)	1
b (0.4)	2	30 × 3.9	μ-Bondapak® C_{18} (10)	E-2	2	UV (210)	9	Ristocetin (6)	—	2

Note: E-1 = Acetonitrile + 0.05 *M* KH_2PO_4, pH 6.
125 : 875

E-2 = Acetonitrile + 0.05 *M* phosphate buffer, pH 6.
9 : 91

REFERENCES

1. **Kirchmeier, R. L. and Upton, R. P.,** *Anal. Chem.*, 50, 349, 1978.
2. **Uhl, J. R. and Anhalt, J. P.,** *Therap. Drug. Monitor.*, 1, 75, 1979.

VERAPAMIL

Gas Chromatography

Specimen (mℓ)	S	Column m × mm	Packing (mesh)	Oven temp (°C)	Gas (mℓ/min)	Det.	RT (min)	Internal standard (RT)	Deriv.	Other compounds (RT)	Ref.
B (1)	1	1.1 × 2.2	3% Dexsil®-300 Supelcon (100/120)	295	Helium (30)	MS-EI	6	[^{13}C,$^{2}H_2$]-Verapamil (6)	a	—	1
B (2)	1	1.8 × 2	3% Dexsil®-300 Chromosorb® W (100/120)	270	Nitrogen (30)	FID	7.3	—	—	b	2

High-Pressure Liquid Chromatography

Specimen (mℓ)	S	Column cm × mm	Packing (μm)	Elution	Flow rate (mℓ/min)	Det. (nm)	RT (min)	Internal standard (RT)	Other compounds (RT)	Ref.
B (0.1—1)	1	25 × 2	MicroPak®[c] MCH-10 (10)	E-1	1	Fl (Ex: 203, Fl: 320)	7.8	d (6.2)	Mono-N-dealkylated metabolite (2.7) Di-N-dealkylated metabolite (3.2)	3

Note: E-1 = Methanol + water + 2-propane sulfonic acid sodium salt + acetic acid, pH 3.9.
156 : 100 : 0.01 moles/ℓ : 1.5 mℓ/ℓ

[a] Plasma extracts are treated with silylating agent to separate interfering peaks. Verapamil does not form a derivative.
[b] The metabolites were analyzed under different conditions.
[c] Column temperature = 30°C.
[d] Lower homolog.

VERAPAMIL (continued)

REFERENCES

1. Spiegelhalder, B. and Eichelbaum, M., *Arzneim-Forsch.*, 27, 94, 1977.
2. McAllister, R. G., Jr., Tan, T. G., and Bourne, D. W. A., *J. Pharm. Sci.*, 68, 574, 1979.
3. Harapat, S. R. and Kates, R. E., *J. Chromatogr.*, 170, 385, 1979.

VIDARABINE

Thin-Layer Chromatography

Specimen (mℓ)	S	Plate (manufacturer)	Layer (mm)	Solvent	Post-separation treatment	Det. (nm)	R_f	Internal standard (R_f)	Other compounds (R_f)	Ref.
D	3	20 × 20 cm (Merck)	Silica gel F_{254} (NA)	S-1	—	Visual (254)	0.6	—	Adenine (0.5)	1

Note: S-1 = Ethanol + acetic acid.
98 : 2

REFERENCE

1. **Hong, W.-H. and Szulczewski, D. H.**, *J. Pharm. Sci.*, 68, 499, 1979.

VINCAMINE

Gas Chromatography

Specimen (mℓ)	S	Column m × mm	Packing (mesh)	Oven temp (°C)	Gas (mℓ/min)	Det.	RT (min)	Internal standard (RT)	Deriv.	Other compounds (RT)	Ref.
B (0.5)	2	2 × 2 (Steel)	3% OV-1 Gas Chrom® Q (100/120)	260	Nitrogen (35)	NPD	3	Diazepam (1.6)	—	—	1
B (1—2)	1	1.5 × 2	1% SE-30 Chromosorb® W (100/120)	210	Helium (25)	MS-EI	8	—	Trimethylsilyl	—	2
B (2)	1	1 × 2	1% OV-1 Gas Chrom® Q (80/100)	235	Nitrgen (20)	NPD	4	Quinine (2.8)	Trimethylsilyl	—	3

Thin-Layer Chromatography

Specimen (mℓ)	S	Plate (manufactuer)	Layer (mm)	Solvent	Post-separation treatment	Det. (nm)	R_f	Internal standard (R_f)	Other compounds (R_f)	Ref.
B (1—5)	2	20 × 20 cm (Merck)	Silica gel G (0.25)	S-1	—	UV Reflectance (280)	0.53	Strychnine (0.25)	Sulpiride[a] (0.45) Naftazone (0.63)	4

Note: S-1 = Chloroform + methanol.
90 : 10

[a] A different solvent is used for developing the plate.

REFERENCES

1. **Kinsun, H. and Moulin, M. A.,** *J. Chromatogr.*, 144, 123, 1977.

2. Hoppen, H.-O., Heuer, R., and Siedel, G., *Biomed. Mass Spectrom.*, 5, 133, 1978.
3. Laufen, H., Juhran, W., Fleissig, W., Götz, R., Scharpf, F., and Bartsch, G., *Arzneim. Forsch.*, 27, 1255, 1977.
4. Bressolle, F., Bres, J., Brun, S., and Rechencq, E., *J. Chromatogr.*, 174, 421, 1979.

VITAMIN A

Gas Chromatography

Specimen (mℓ)	S	Column m × mm	Packing (mesh)	Oven temp (°C)	Gas (mℓ/min)	Det.	RT (min)	Internal standard (RT)	Deriv.	Other compounds (RT)	Ref.
D	3	1.8 × NA	5% OV-1 Chromosorb® W (80/100)	230	Helium (40)	FID	3.5—7	Methylhelptadecanoate (5)	Catalytic hydrogenation	—	1
B (2)	1	1.8 × 2	1% FFAP Gas Chrom® Q (100/120)	190	Helium (30)	MS-EI	—	Methyl-d_3-retinoate (11)	Methyl[a]	Vitamin A acid (11)	2

High-Pressure Liquid Chromatography

Specimen (mℓ)	S	Column cm × mm	Packing (μm)	Elution	Flow rate (mℓ/min)	Det. (nm)	RT (min)	Internal standard (RT)	Other compounds (RT)	Ref.
B (0.2)	2	15 × 3.2	R Sil C_{18} (10)	E-1	1.0	UV (330)	1.0	Retinylpropionate (1.4)	Retinyllinoleate (8.5) Retinylpalmitate (11) Retinylstearate (15.5)	3
B, U (1)	1	25 × 4.6	Partisil® (10)	E-2	2.2	Absorbance (365)	—	—	13-*cis*-Vitamin A acid (2.7) *trans*-Vitamin A acid (3.5)	4
D	3	25 × 7.9	Zorbax® SIL (NA)	E-3	2	UV (254)	—	2,6-Di-tert-butyl-*p*-cresol (5)	13-*cis*-Retinal (11.8) 11-*cis*-Retinal (13.5) 9-*cis*-Retinal (14.9)	5

D	3	25 × 3	Silica gel Si60 (5)	E-4	1.05	UV (320)	ed	—	all-*trans*-Retinal (19.4) 13-*cis*-Retinylpalmitate (3.2) 11-*cis*-Retinylpalmitate (3.4) 9-*cis*-Retinylpalmitate (3.9) all-*trans*-Retinylpalmite (4.5)	6
D	3	25 × 3	Micropak® Si-5 (5)	E-5	1	Absorbance (360)	—	—	syn-Alltrans-Retinaloxine (2.2) anit-Alltrans-Retinaloxine (4.8) syn-11-*cis*-Retinaloxine (1.8) anti-11-*cis*-Retinaloxine (3.8)	7
D	3	30 × 4	μ-Bondapak® C_{18} (10)	E-6	2	UV (313)	22	—	Vitamin A acid (11) Retinal (25) Retinylpalmitate (51)	8

Thin-Layer Chromatography

Specimen (mℓ)	S	Plate (manufacturer)	Layer (mm)	Solvent	Post-separation treatment	Det. (nm)	R_f	Internal standard (R_f)	Other compounds (R_f)	Ref.
D	3	5 × 20 cm (Merck)	Silica gel 60 F_{254}[b] (0.25)	S-1[c]	—	Visual (366)	0.26	—	Vitamin A acid (0.21) Retinal (0.40) Retinolacetate (0.54) Retinolpalmitate (0.71)	9
D	3	20 × 20 cm	Silica gel G	S-2	—	Fl	0.96	—	Vitamin B_1 (0.63)	10

Vitamin A (continued)

Thin-Layer Chromatography (continued)

Specimen (mℓ)	S	Plate (manufacturer)	Layer (mm)	Solvent	Post-separation treatment	Det. (nm)	R_f	Internal standard (R_f)	Other compounds (R_f)	Ref.
		(Laboratory)	(0.25)			(Ex: 254, Fl: 439)			Vitamin B_2 (0.20)[d] Vitamin C (0.42) Vitamin D_3 (0.85)	

Note: E-1 = Methanol.

E-2 = Methylene chloride + acetic acid.
99.5 : 0.5

E-3 = Diethyl ether + *n*-hexane.
12 : 88

E-4 = *n*-Hexane + dioxane.
99.1 : 0.1

E-5 = Hexane + dioxane.
95 : 5

E-6 = 0.01 *M* Sodium acetate in methanol + water.
80 : 20

S-1 = Acetone + petroleum ether.
18 : 82

S-2 = Acetone + methanol + benzene.
1 : 2 : 8

a. With diazomethane.
b. The plates are sprayed with diethyl ether containing 50 μg/mℓ of butylated hydroxytoluene prior to use.
c. The plates are developed in the dark.
d. Emission wavelength for Vitamin B_2 = 532 nm.

REFERENCES

1. Fenton, T. W., Vogtmann, H., and Clandinin, D. R., *J. Chromatogr.*, 77, 410, 1973.
2. de Ruyter, M. G., Lambert, W. E., and de Leenheer, A. P., *Anal. Biochem.*, 98, 402, 1979.
3. de Ruyter, M. G. M. and de Leenheer, A. P., *Clin. Chem.*, 24, 1920, 1978.
4. Puglisi, C. V. and de Silva, J. A. F., *J. Chromatogr.*, 152, 421, 1978.
5. Tsukida, K., Kodama, A., and Ito, M., *J. Chromatogr.*, 134, 331, 1977.
6. Paanaker, J. E. and Groenendijk, G. W. T., *J. Chromatogr.*, 168, 125, 1979.
7. Groenendijk, G. W. T., de Grip, W. J., and Daemen, F. J. M., *Anal. Biochem.*, 99, 304, 1979.
8. McCormick, A. M., Napoli, J. L., and Deluca, H. F., *Anal. Biochem.*, 86, 25, 1978.
9. Fung, Y. K., Rahwan, R. G., and Sams, R. A., *J. Chromatogr.*, 147, 528, 1978.
10. Kouimtzis, Th. A. and Papadoyannis, I. N., *Microchim. Acta*, 1, 145, 1979.

VITAMIN B_{12}

High-Pressure Liquid Chromatography

Specimen (mℓ)	S	Column cm × mm	Packing (μm)	Elution	Flow rate (mℓ/min)	Det. (nm)	RT (min)	Internal standard (RT)	Other compounds (RT)	Ref.
D	3	30 × 3.9	μ-Bondapak® C_{18} (10)	E-1 (Gradient)	1.8	UV (254)	6.5	—	Hydroxycobalamin (3) Deoxyadenosylcobalamin (9) Methylcobalamin (11.5)	1

Note: i. 27% Aqueous methanol.
ii. 95% Aqueous methanol.

REFERENCE

1. Frenkel, E. P., Kitchens, R. L., and Prough, R., *J. Chromatogr.*, 174, 393, 1979.

VITAMIN C

High-Pressure Liquid Chromatography

Specimen (mℓ)	S	Column cm × mm	Packing (μm)	Elution	Flow rate (mℓ/min)	Det. (nm)	RT (min)	Internal standard (RT)	Other compounds (RT)	Ref.
U (0.5) D	2	50 × 2	Zipax®[a] SAX (NA)	E-1	0.33	Electro-chemical	6	—	—	1
D	3	30 × 4	μ-Bondapak® C_{18} (10)	E-2	3.0	UV (254)	5.2	—	—	2
U (0.005)	1	30 × 4	μ-Bondapak® C_{18} (10)	E-3	3.0	UV (254)	0.9	—	—	3

Thin-Layer Chromatography

Specimen (mℓ)	S	Plate (manufacturer)	Layer (mm)	Solvent	Post-separation treatment	Det. (nm)	R_f	Internal standard (R_f)	Other compounds (R_f)	Ref.
D	3	Glass-fiber paper (Gelman)	—	S-1	Charring with conc. H_2SO_4	Visual	0.62	—	—	4

Note: E-1 = 0.05 *M* Acetate buffer, pH 4.75.

E-2 = Methanol + 2 m*M* tridecylammonium formate, pH 5.
300 : 300

E-3 = 0.8% Metaphosphoric acid.

S-1 = Acetonitrile + butyronitrile + ethanol + water.
60 : 30 : 12 : 2

[a] Column temperature = 25°C.

VITAMIN C (continued)

REFERENCES

1. **Pachla, L. A. and Kissinger, P. T.,** *Anal. Chem.*, 48, 364, 1976.
2. **Sood, S. P., Sartori, L. E., Wittmer, D. P., and Haney, W. G.,** *Anal. Chem.*, 48, 796, 1976.
3. **Wagner, E. S., Lindley, B., and Coffin, R. D.,** *J. Chromatogr.*, 163, 225, 1979.
4. **Hornig, D.,** *J. Chromatogr.*, 71, 169, 1972.

VITAMIN D

Gas Chromatography

Specimen (mℓ)	S	Column m × mm	Packing (mesh)	Oven temp (°C)	Gas (mℓ/min)	Det.	RT (min)	Internal standard (RT)	Deriv.	Other compounds (RT)	Ref.
B (2.5)	1	2.5 × 2	1.5% SE-30 Chromosorb® W (80/100)	250	Helium (15)	MS-EI	—	25-Hydroxy-[26^2H_3] vitamin D_3 (2,4)[a]	Trimethylsilyl	25-Hydroxy-vitamin D_3 (2,4)[a]	1
B (5)	1	2×2	1% FFAP Gas Chrom® Q (100/120)	215	Helium (NA)	MS-EI	D_3 = 7	Dihydrotachysterol$_2$ (4)	Heptafluorobutyryl	—	2

High-Pressure Liquid Chromatography

Specimen (mℓ)	S	Column cm × mm	Packing (μm)	Elution	Flow rate (mℓ/min)	Det. (nm)	RT (min)	Internal standard (RT)	Other compounds (RT)	Ref.
B (25)	2	25 × 2.1	Zorbax®-ODS (5)	E-1	0.5	UV (254)	—	—	25-Hydroxy-vitamin D_3 (6)	3
D	3	30 × 4	μ-Porasil® (10)	E-2	1	UV (254)	D_3 = 17	*p*-Dimethylaminobenzaldehyde (13)	*trans*-vitamin D_3 (10) Tachysterol$_3$ (15) 7-Dehydrocholesterol (24)	4
B (4)	2	30 × 4	μ-Porasil® (10)	E-3	2	UV (254)	—	—	25-Hydroxy-vitamin D_2 (7.2) 25-Hydroxy-vitamin D_3 (9)	5
B (1)	2	30 × 4	μ-Porasil® (10)	E-4	0.5	UV (254)	—	—	25-Hydroxy-vitamin D_3 (16) 25-Hydroxy-vitamin D_2 (14)	6

VITAMIN D (continued)

High-Pressure Liquid Chromatography

Specimen (mℓ)	S	Column cm × mm	Packing (μm)	Elution	Flow rate (mℓ/min)	Det. (nm)	RT (min)	Internal standard (RT)	Other compounds (RT)	Ref.
D	3	30 × 4	μ-Porasil® (10)	E-5	2	UV (254)	D_2 = 8 D_3 = 8	—	24,25-Dihydroxy-vitamin D_3 (52) Lumisterol$_3$ (5.5) Previtamin D_2 (5) Previtamin D_3 (5) Dihydrocholes-terol (10)	7
B (2)	2	22 × 6.2	Zorbax® Sil (NA)	E-6	1.5	UV (254)	D_2 = 5.5[b] D_3 = 5.5	—	25-Hydroxy-vita-min D_2 (12) 25-Hydroxy-vita-min D_3 (13)	8
Fish-liver oil	2	25 × 3	LiChrosorb® RP-18 (10)	E-7	1	UV (263)	D_2 = 7.5[c] D_3 = 8.4	—	Previtamin D_2 6.6) Previtamin D_3 (7.3)	9
D	3	25 × 4.6	Partisil® 10 (10)	E-8	0.8	UV (254)	D_2 = 20	4-Hydroxybi-phenyl (25)	Vitamin A acetate (8)	10
B (0.8)	2	25 × 4	μ-Porasil® (10)	E-9	1	UV (254)	—	—	25-Hydroxy-vita-min D_3 (16)	11

Note: E-1 = Acetonitrile + methanol + water.
90 : 5 : 5

E-2 = Chloroform + *n*-hexane + tetrahydrofuran.
70 : 30 : 1

E-3 = 2-propanol + hexane.
2.5 : 97.5

E-4 = Ethanol + hexane.
5 : 95

E-5 = Petroleum ether + 1,2-dichloroethane + tetrahydrofuran.
85 : 8 : 7

E-6 = Isopropanol + hexane.
5.5 : 94.5

E-7 = Methanol + water.
95 : 5

E-8 = Isopropanol + cyclohexane.
1.25 : 98.75

E-9 = Isopropanol + methylene chloride.
11 : 49

[a] Two peaks correspond to the pyro- and isopyro-forms of trimethylsilyl ether of 25-hydroxy-vitamin D_3.
[b] Fractions were collected and reanalyzed by reverse-phase liquid chromatography using Zorbax® ODS column and 98.5% methanol-water as the mobile phase.
[c] Extracts of liver were chromatographed by adsorption chromatography using a Partisil® 10 column and 4% isopropanol in hexane as the mobile phase. Fractions were collected and rechromatographed under the conditions described.

REFERENCES

1. **Björkhem, I. and Holmberg, I.,** *Clin. Chim. Acta,* 68, 215, 1976.
2. **de Leenheer, A. P. and Cruyl, A. A.,** *Anal. Biochem.,* 91, 293, 1978.
3. **Koshy, K. T. and Vander Slik, A. L.,** *Anal. Biochem.,* 74, 282, 1976; 85, 283, 1978.
4. **Tartivita, K. A., Sciarello, J. P., and Rudy, B. C.,** *J. Pharm. Sci.,* 65, 1024, 1976.
5. **Eisman, J. A., Shepard, R. M., and Deluca, H. F.,** *Anal. Biochem.,* 80, 298, 1977.
6. **Gilbertson, T. J. and Stryd, R. P.,** *Clin. Chem.,* 23, 1700, 1977; 24, 927, 1978.
7. **Vanhaelen-Fastre, R. and Vanhaelen, M.,** *J. Chromatogr.,* 153, 219, 1978.
8. **Jones, G.,** *Clin. Chem.,* 24, 287, 1978.
9. **Egaas, E. and Lambertsen, G.,** *Int. J. Vitam. Nutr. Res.,* 49, 35, 1979.
10. **Mackay, C., Tillman, J., and Burns, D. T.,** *Analyst (London),* 104, 626, 1979.
11. **Schaefer, P. C. and Goldsmith, R. S.,** *J. Lab. Clin. Med.,* 91, 104, 1978.

VITAMIN E

Gas Chromatography

Specimen (mℓ)	S	Column m × mm	Packing (mesh)	Oven temp (°C)	Gas (mℓ/min)	Det.	RT (min)	Internal standard (RT)	Deriv.	Other compounds (RT)	Ref.
B (0.1)	2	4.5 × NA	0.5% Apiezon L Gas Chrom® Q (100/120)	250	Helium (25)	FID[c]	22	5,7-Dimethylto-col (20)	Trimethyl-silyl	γ-Tocopherol (15)	1
B (3)	2	1.8 × NA	3% OV-1 Supelcoport (80/100)	240	Helium (80)	FID[c]	19	5,7-Dimethylto-col (17)	Trimethyl-silyl	β-Tocopherol (13)	2
B (1)	2	32 × 0.25	PZ-176[a]	Temp. Progr.	Nitrogen[b]	FID	56	—	Trimethyl-silyl	δ-Tocopherol (46) β-Tocopherol (49) γ-Tocopherol (49.3)	3
B (0.5)	2	2 × 4	3% SE-30 Gas Chrom® Q (100/120)	242	Nitrogen (70)	FID[c]	21.5	—	Acetyl	—	4

High-Pressure Liquid Chromatography

Specimen (mℓ)	S	Column cm × mm	Packing (μm)	Elution	Flow rate (mℓ/min)	Det. (nm)	RT (min)	Internal standard (RT)	Other compounds (RT)	Ref.
D	3	25 × 2.1	MicroPak® CH10 (10)	E-1	1	UV (285)	4.9	—	α-Tocopheryl ace-tate (7) δ-Tocopheryl ace-tate (5.2)	5
D	3	25 × 4.6	Partisil® PX210 (10)	E-2	2.39	Fl (Ex: 295, Fl: 320)	NA	—	d	6
B (0.2)	2	25 × 4.6[e]	LiChrosorb® RP-18 (10)	E-3	2	UV (292)	5.4	Tocol (3.8)	β,γ-Tocopherol (4.9)	7

Animal feed	2	30 × 3.9	μ-Bondapak® C_{18} (10)	E-4	3	Fl (Ex: 296, Fl: 330)	5.5[f]	—	—	8

Note: E-1 = 3% Water in methanol.

E-2 = 0.3% Methanol in hexane.

E-3 = Methanol.

E-4 = 5% Water in methanol.

[a] Open tubular-glass capillary column.
[b] Initial column pressure = 5 psi.
[c] Plasma extracts were purified by thin-layer chromatography prior to analysis.
[d] In this system different isomerc tocopherols are separated.
[e] Column temperature = 40°C.
[f] Feed extracts are saponified to convert any tocopherol acetate to free tocopherol.

REFERENCES

1. Lehmann, J. and Slover, H. T., *Lipids,* 6, 35, 1971.
2. Lovelady, H. G., *J. Chromatogr.,* 85, 81, 1973.
3. Lin, S.-N. and Horning, E. C., *J. Chromatogr.,* 112, 465, 1975.
4. Chiarotti, M. and Giusti, G. V., *J. Chromatogr.,* 147, 481, 1978.
5. Eriksson, T. and Sörensen, B., *Acta Pharm. Suec.,* 14, 475, 1977.
6. Vatassery, G. T., Maynard, V. R., and Hagen, D. F., *J. Chromatogr.,* 161, 299, 1978.
7. de Leenheer, A. P., de Bevere, V. O., Cryl, A. A., and Claeys, A. E., *Clin. Chem.,* 24, 585, 1978; 25, 425, 1978.
8. McMurray, C. H. and Blanchflower, W. J., *J. Chromatogr.,* 176, 488, 1979.

VITAMIN K_1

Gas Chromatography

Specimen (mℓ)	S	Column m × mm	Packing (mesh)	Oven temp (°C)	Gas (mℓ/min)	Det.	RT (min)	Internal standard (RT)	Deriv.	Other compounds (RT)	Ref.
D	3	3 × 2	5% XE-60 Chromosorb® G (80/100)	190	NA (30)	FID	—	1,4-Naphthaquinone (9.5)	—	Menadione (12.5)	1
B (0.2—0.8)	1	1.9 × 2	3% OV-17 Anakam Q (90/100)	302	Nitrogen (80)	ECD	4.6	Vitamin K_2 (20) (7)	—	Vitamin $K_{1-2,3}$ epoxide (4.1)	2

Thin-Layer Chromatography

Specimen (mℓ)	S	Plate (manufacturer)	Layer (mm)	Solvent	Post-separation treatment	Det. (nm)	R_f	Internal standard (R_f)	Other compounds (R_f)	Ref.
Elemental diet	2	20 × 20 cm (Merck)	Silica gel-60 F-254 (NA)	S_1	—	UV Reflectance (270)	0.40	—	—	3

Note: S-1 = Toluene.

REFERENCES

1. **Castello, G., Bruschi, E., and Ghelli, G.,** *J. Chromatogr.,* 139, 195, 1977.
2. **Bechtold, H. and Jähnchen, E.,** *J. Chromatogr.,* 164, 85, 1979.
3. **Horn, G. B.,** *J. Pharm. Sci.,* 67, 834, 1978.

WARFARIN

Gas Chromatography

Specimen (mℓ)	S	Column m × mm	Packing (mesh)	Oven temp (°C)	Gas (mℓ/min)	Det.	RT (min)	Internal standard (RT)	Deriv.	Other compounds (RT)	Ref.
B (1)	1	0.6 × 3	1% OV-17 Gas Chrom® Q (80/100)	240	Nitrogen (75)	ECD	3.6	*p*-Chlorowarfarin (6.1)	Pentafluorobenzyl	—	1
B (2)	1	1.8 × 2.5	5% OV-7 Chromosorb® W (80/100)	260	Nitrogen (63)	FID	9.8	Phenylbutazone (7.7)[a]	Methyl[b]	—	2
B (2)	1	1.8 × 2.5	3.8% UCW-98 Gas Chrom® W (80/100)	270	Helium (60)	FID	2	Papaverine (3.1)	—	5-Hyroxywarfarin (0.6) 7-Hydroxywarfarin (0.7) 8-Hydroxywarfarin (0.8)	3
B (2)	1	1.8 × 4	5% OV-7 Chromosorb® W (80/100)	250	Helium (125)	FID	8.8	Phenylbutazone (7)[a]	Methyl[c]	—	4

High-Pressure Liquid Chromatography

Specimen (mℓ)	S	Column cm × mm	Packing (μm)	Elution	Flow rate (mℓ/min)	Det. (nm)	RT (min)	Internal standard (RT)	Other compounds (RT)	Ref.
B (NA)	1	100 × 2.1	Permaphase (NA)	E-1	0.75	UV (254)	8	—	7-Hydroxywarfarin (4)	5
B, U (10) Tissue	2	100 × 1.5	Corasil II (NA)	E-2	1	UV (270)	2	—	—	6
B (0.2—1)	1	25 × 2.2	MicroPak® CH-10 (10)	E-3	1	UV (308)	3.3	*p*-Chlorowarfarin (6)	—	7
B (1.3)	1	30 × 4	μ-Bondapak® C_{18}	E-4	2	UV (313)	15	—	4′-Hydroxywar-	8

WARFARIN (continued)

High-Pressure Liquid Chromatography

Specimen (mℓ)	S	Column cm × mm	Packing (μm)	Elution	Flow rate (mℓ/min)	Det. (nm)	RT (min)	Internal standard (RT)	Other compounds (RT)	Ref.
			(10)						farin (5) 6-Hydroxywarfarin (6) 8-Hydroxywarfarin (7.5) 7-Hydroxywarfarin (8.2)	
B (1—2)	1	25 × 2.2	MicroPak® CH-10 (10)	E-5	0.83	UV (305)	3	Methylated warfarin (5)	7-Hydroxywarfarin (2)	9
B (0.5—1)	1	30 × 4	μ-Bondapak® C_{18} (10)	E-6	1	UV (313)	5.5	*p*-Chlorowarfarin (6.7)	—	10

Thin-Layer Chromatography

Specimen (mℓ)	S	Plate (manufacturer)	Layer (mm)	Solvent	Post-separation treatment	Det. (nm)	R_f	Internal standard (R_f)	Other compounds (R_f)	Ref.
D	3	10 × 20 cm (Laboratory)	Silica gel F_{254}[d] (0.25)	S-1	—	Visual (254,365)	0.34	—	Dicoumarol (0.15) Phenprocoumon (0.56) Acenocoumarin (0.23)	11
D	3	10 × 10 cm (Merck)	Silica gel F_{254}[e] (NA)	S-2	—	Visual (254)	0.28	—	4-Hydroxycoumarin (0.09) Phenprocoumon 0.49) Clocoumarol (0.53)	12

Acenocoumarin (0.20)
Bishydroxycoumarin (0.45)
Ethylbiscoumacetate (0.09)

Note: E-1 = Dioxane + water, pH 4.
10 : 70

E-2 = Isopropyl alcohol + isooctane.
2 98

E-3 = Methanol + 0.5% acetic acid.
50 : 50

E-4 = Acetonitrile + 1.5% acetic acid, pH 4.7.
31 : 69

E-5 = Dioxane + water, pH 4.2.
40 : 60

E-6 = Methanol + acetic acid + water.
75 : 0.5 : 25

S-1 = Benzene + diethyl ether (plate buffered at pH 2.0).
2 : 1

S-2 = Benzene + carbon tetrachloride + dioxane + acetic acid.
50 : 40 : 10 : 1

[a] Phenylbutazone gives two peaks. However, the peak with a higher retention time is used for quantitation.
[b] With diazomethane.
[c] On column methylation with trimethylanilinium hydroxide.
[d] Silica gel is buffered at pH 2.0 or pH 3.6 or pH 5.0.
[e] High-performance thin-layer plates.

REFERENCES

1. **Kaiser, D. G. and Martin, R. S.,** *J. Pharm. Sci.,* 63, 1579, 1974.

WARFARIN (continued)

2. Midha, K. K., McGilveray, I. J., and Cooper, J. K., *J. Pharm. Sci.*, 63, 1725, 1974.
3. Hanna, S., Rosen, M., Eisenberger, P., Rasero, L., and Lachman, L., *J. Pharm. Sci.*, 67, 84, 1978.
4. Loomis, C. W. and Racz, W. J., *Anal. Chim. Acta*, 106, 155, 1979.
5. Vessel, E. S. and Shively, C. A., *Science*, 184, 466, 1974.
6. Mundy, D. E., Quick, M. P., and Machin, A. F., *J. Chromatogr.*, 121, 335, 1976.
7. Bjornsson, T. D., Blaschke, T. F., and Meffin, P. J., *J. Pharm. Sci.*, 66, 142, 1977.
8. Fasco, M. J., Piper, L. J., and Kaminsky, L. S., *J. Chromatogr.*, 131, 365, 1977; *J. Liq. Chromatogr.*, 2, 565, 1979.
9. Wong, L. T., Solomonraj, G., and Thomas, B. H., *J. Chromatogr.*, 135, 149, 1977.
10. Forman, W. B. and Shlaes, J., *J. Chromatogr.*, 146, 522, 1978.
11. Lau-Cam, C. A., *J. Chromatogr.*, 151, 391, 1978.
12. Vanhaelen-Fastre, R. and Vanhaelen, M., *J. Chromatogr.*, 129, 397, 1976.

YC-93

Gas Chromatography

Specimen (mℓ)	S	Column m × mm	Packing (mesh)	Oven temp (°C)	Gas (mℓ/min)	Det.	RT (min)	Internal standard (RT)	Deriv.	Other compounds (RT)	Ref.
B	1	0.9 × 1.8	3% OV-1 Chromosorb® W (80/100)	260	Methane 5 + Argon 95 (40)	ECD	3.3	a (4.4)	Oxidation with ni-trous acid	—	1

[a] 2,6-Dimethyl-4-(3-nitrophenyl)-1,4-dihydropyridine-3,5-dicarboxylic acid-3-[3-*N*-benyl-*N*-methyl-amino)]-propyl ester-5-isopropyl ester.

REFERENCE

1. **Higuchi, S., Sasaki, H., and Sado, T.,** *J. Chromatogr.*, 110, 301, 1975.

ZIMELIDINE

Gas Chromatography

Specimen (mℓ)	S	Column m × mm	Packing (mesh)	Oven temp (°C)	Gas (mℓ/min)	Det.	RT (min)	Internal standard (RT)	Deriv.	Other compounds (RT)	Ref.
B (2)	2	0.9 × 4	1% OV-17 Chromosorb® 75 (100/120)	240	Argon 90 + Methane (10) (50)	ECD	2.6	CPK191[a] (2)	—	Norzimelidine (3.3)	1
B (1)	2	2 × 2	3% OV-17 Supelcoport (80/100)	255	Nitrogen (30)	NPD	3.4	Propyl analog (5.6)	—	Norzimelidine (4.1)	2

High-Pressure Liquid Chromatography

Specimen (mℓ)	S	Column cm × mm	Packing (μm)	Elution	Flow rate (mℓ/min)	Det. (nm)	RT (min)	Internal standard (RT)	Other compounds (RT)	Ref.
B (1)	2	12 × 4.5	Partisil®[b] (5)	E-1	0.83	UV (258)	2.5, 3.5[c]	—	Norzimelidine (6,8)[c]	3
B (1), U (NA) Tissue	2	15 × 4	Partisil® (6)	E-2	0.75	UV (262)	5.8	Chlorphenira-mine (4.2)	Norzimelidine (6.5, 8)[c]	4

Note: E-1 = Methanol + 0.1 *M* ammonium nitrate.
100 : 5

E-2 = Dichloromethane + *n*-butanol saturated with the stationary phase (0.2 *M* $HClO_4$ + 0.8 *M* $NaClO_4$).
89 : 11

[a] *N*-Methyl-3-(4-bromophenyl)-3-phenylallylamine.
[b] Column temperature = 30°C.
[c] Retention times of geometric isomers.

REFERENCES

1. Larsen, N.-E. and Marinelli, K., *J. Chromatogr.*, 156, 335, 1978.
2. Hogberg, K., Lindgren, J.-E., Högberg, T., and Ulff, B., *Acta Pharm. Suec.*, 16, 299, 1979.
3. Emanuelsson, B. and Moore, R. G., *J. Chromatogr.*, 146, 113, 1978.
4. Westerlund, D., Nilsson, L. B., and Jaksch, Y., *J. Liq. Chromatogr.*, 2, 373, 1979.

APPENDIX

MOLECULAR WEIGHT OF DRUGS IN TABLES

Only drugs which have been described in the tables of this handbook and are also listed in the Merck Index (Martha Windholz, Ed., 9th ed.) are included in this appendix.

The mol. wt. is of free acid or of free base.

The generic name of the drug as it appears on the title of the table is given first. Some of the synonyms of the generic name of a drug are given in parentheses. In some cases, a few commonly used proprietary names for a drug are also given.

Drug Name	Molecular formula	Molecular weight
Acebutolol	$C_{18}H_{28}N_2O_4$	336.43
Acenocoumarin	$C_{19}H_{15}NO_6$	353.32
Sintrom®		
Acephylline	$C_9H_{10}N_4O_4$	238.20
(Acepiphylline, Acefylline, 7-Theophyllineacetic acid)		
Etaphylline		
Acetaminophen	$C_8H_9NO_2$	151.16
(Paracetamol)		
Acamol, Calpol, Tempra®, Tylenol®		
Acetanilide	C_8H_9NO	135.16
(Antifebrin)		
Acetazolamide	$C_4H_6N_4O_3S_2$	222.25
Acetamox, Cidamex, Diamox®, Nephramid		
Acetohexamide	$C_{15}H_{20}N_2O_4S$	324.42
Dimelor, Dymelor®, Dimelin, Ordimel		
Acetylmethadol	$C_{23}H_{31}NO$	353.49
(Methadylacetate)		
Acemethadone, Amidolacetate, Race-Acetylmethadol		
Acetylsalicylic Acid	$C_9H_8O_4$	180.15
(Aspirin)		
Acentrine, Acetophen, Acetosal, Acetosalic acid, Acetosalin		
Adriamycin	$C_{27}H_{29}NO_{11}$	543.54
(Doxorubicin)		
Alclofenac	$C_{11}H_{11}ClO_3$	226.66
Argun, Mervan, Neoston, Prinalgin®		
Allobarbital	$C_{10}H_{12}N_2O_3$	208.21
Malilum, Diadol, Dial®		
Allopurinol	$C_5H_4N_4O$	136.11
Zyloprim®, Belminol, Bloxanth, Epidropal, Foligan		
Alprenolol	$C_{15}H_{23}NO_2$	249.34
Amikacin	$C_{22}H_{43}N_5O_{13}$	585.62
ε-Aminocaproic Acid	$C_6H_{13}NO_2$	131.17
(6-Aminohexanoic Acid)		
Amicar, Capramol, Epsamon, Epsikapron		
Aminopyrine	$C_{13}H_{17}N_3O$	231.29
Amidopyrine, Aminophenazone, Dipirin		
Amitriptyline	$C_{20}H_{23}N$	277.39
Elavil®, Deprox, Levate, Meravil, Novotriptyn		
Amobarbital	$C_{11}H_{18}N_2O_3$	226.27
Amytal®, Isobec		
Amoxicillin	$C_{16}H_{19}N_3O_5S$	365.41
Amphetamine	$C_9H_{13}N$	135.20
(Dexamphetamine)		
Dexedrine		

Appendix (Continued)

MOLECULAR WEIGHT OF DRUGS IN TABLES

Drug Name	Molecular formula	Molecular weight
Amphotericin B Fungizone®	$C_{47}H_{73}NO_{17}$	924.11
Amipcillin	$C_{16}H_{19}N_3O_4S$	349.42
Amygdalin	$C_{20}H_{27}NO_{11}$	457.42
Antipyrine (Phenazone) Analgesine	$C_{11}H_{12}N_2O$	188.22
Apomorphine	$C_{17}H_{17}NO_2$	267.31
Aprinidine	$C_{22}H_{30}N_2$	322.49
Atenolol Tenormin®	$C_{14}H_{22}N_2O_3$	266.34
Atropine (Hyoscyamine)	$C_{17}H_{23}NO_3$	289.38
Azapropazone (Apazone) Rheumox, Sinnamin, Cinnamin	$C_{16}H_{20}N_4O_2$	300.37
6-Azauridine-Triacetate (Azaribine) Triazure	$C_{14}H_{17}N_3O_9$	
Bacitracin Baciquent, Bacitin	$C_{66}H_{103}N_{17}O_{16}S$	1411
Baclofen Liorseal®	$C_{10}H_{12}ClNO_2$	213.67
Barbital Veronal, Malonal	$C_8H_{12}N_2O_3$	184.19
Benactyzine Actozine, Levol, Suavitil®	$C_{20}H_{25}NO_3$	327.41
Bendroflumethiazide (Bendrofluazide, Benzydroflumethiazide) Naturetin®, Aprinox, Benuron	$C_{15}H_{14}F_3N_3O_4S_2$	421.41
Benorylate Benoral, Benortan	$C_{17}H_{15}NO_5$	313.32
Benzarone Fragivix, Fragivil	$C_{17}H_{14}O_3$	266.28
Benzoyl Peroxide Benoxyl, Persadox, Lucidol	$C_{14}H_{10}O_4$	242.22
Benzthiazide ExNa®	$C_{15}H_{14}ClN_3O_4S_3$	431.96
Berberine	$C_{20}H_{18}NO_4$	336.37
Betahistine	$C_8H_{12}N_2$	136.19
Bezitramide Burgodin	$C_{31}H_{32}N_4O_2$	492.63
Bleomycins	Mixture	
Bromazepam RO 5-3350 Lectopam Lexotanil®	$C_{14}H_{10}BrN_3O$	316.16
Bromhexine Bisolvon	$C_{14}H_{20}Br_2N_2$	376.1
Bromocriptine	$C_{32}H_{40}BrN_5O_5$	654.62
Bupivacaine	$C_{18}H_{28}N_2O$	288.43

Appendix (Continued)

MOLECULAR WEIGHT OF DRUGS IN TABLES

Drug Name	Molecular formula	Molecular weight
Butabarbital	$C_{10}H_{16}N_2O_3$	212.23
(Secbutabarbital)		
Buta-Barb, Day-Barb, Butisol®		
Butriptyline	$C_{21}H_{27}N$	293.46
Caffeine	$C_8H_{10}N_4O_2$	194.19
No-Doz®		
Camphor	$C_{10}H_{16}O$	152.23
Camylofine	$C_{19}H_{32}N_2O_2$	320.46
Adopan, Spasmocan®		
Cannabidiol	$C_{21}H_{31}O_2$	314.45
Cannabinol	$C_{21}H_{26}O_2$	310.42
Carbamazepine	$C_{15}H_{12}N_2O$	236.26
Finlepsin, Tegretol®		
Carpronium Chloride	$C_8H_{13}ClNO_2$	195.70
Actinamin		
Cefazolin	$C_{14}H_{14}N_8O_4S_3$	454.50
Cefoxitin	$C_{16}H_{17}N_3O_7S_2$	427.46
Cephacetrile	$C_{13}H_{13}N_3O_6S$	339.31
Cephalexin	$C_{16}H_{17}N_3O_4S$	347.40
Cephaloridine	$C_{19}H_{17}N_3O_4S_2$	415.50
Cephalosporin C	$C_{16}H_{21}N_3O_8S$	415.44
Cephalothin	$C_{16}H_{16}N_2O_6S_2$	396.44
Chloral Hydrate	$C_2H_3Cl_3O_2$	165.42
Noctec®, Nycton, Somnos, Lorinal		
Chlorambucil	$C_{14}H_{19}Cl_2NO_2$	304.23
Amboclorin, Leukeran, Linfolysin		
Chloramphenicol	$C_{11}H_{12}Cl_2N_2O_5$	323.14
Chloromycetin®, Chloroptic, Fenicol		
Chlordiazepoxide	$C_{16}H_{14}ClN_3O$	299.75
Librium®, Dymopoxide, Corax, Diapax, Medilium		
Chlorhexidine	$C_{22}H_{30}Cl_2N_{10}$	505.48
Hibitane		
Chlorobutanol	$C_4H_7Cl_3O$	177.47
Chloretone®		
Chloroform	$CHCl_3$	119.39
Chloroprocaine	$C_{13}H_{19}ClN_2O_2$	270.72
Nesacaine®		
Chloroquine	$C_{18}H_{26}ClN_3$	319.89
Aralen®		
Chlorothiazide	$C_7H_6ClN_3O_4S_2$	295.74
Chlorphenesin Carbamate	$C_{10}H_{12}ClNO_4$	245.68
Maolate		
Chlorpheniramine	$C_{16}H_{19}ClN_2$	274.80
Chlor-Tripolon, Histalon, Histaspan, Novopheniram		
Chlorpromazine	$C_{17}H_{19}ClN_2S$	318.88
Chlorprom, Largactil®		
Chlorpropamide	$C_{10}H_{13}ClN_2O_3S$	276.75
Diabinese®, Chloromide, Chloronase, Novopropamide, Stabinol		
Chlortetracycline	$C_{22}H_{23}ClN_2O_8$	478.88
Aureomycin®		
Chlorthalidone	$C_{14}H_{11}ClN_2O_4S$	338.78
Hydroton, Hygroton®		
Chlorzoxazone	$C_7H_4ClNO_2$	169.58
Paraflex		
Choline	$C_5H_{15}NO_2$	121.18

Appendix (Continued)

MOLECULAR WEIGHT OF DRUGS IN TABLES

Drug Name	Molecular formula	Molecular weight
Cimetidine	$C_{10}H_{16}N_6S$	252.34
Tagamet®		
Cinnarizine	$C_{26}H_{28}N_2$	368.50
Dimitron, Glanil, Mitronal		
Citrovorum Factor	$C_{20}H_{23}N_7O_7$	473.44
(Folinic Acid)		
Leucovorin		
Clobazam	$C_{16}H_{13}ClN_2O_3$	300.74
Clofibrate	$C_{12}H_{15}ClO_3$	242.71
(Clofibric Acid Ethyl Ester	$C_{10}H_{11}ClO_3$	214.66)
Atromid-S®		
Clomipramine	$C_{19}H_{23}ClN_2$	314.87
Anafranil®		
Clonazepam	$C_{15}H_{10}ClN_3O_3$	315.72
Rivotril		
Clonidine	$C_9H_9Cl_2N_3$	230.10
Clorazepate	$C_{16}H_{13}ClN_2O_4$	332.74
Clozapine	$C_{18}H_{19}ClN_4$	326.83
Leponex		
Cocaine	$C_{17}H_{21}NO_4$	303.35
Codeine	$C_{18}H_{21}NO_3$	299.36
Colchicine	$C_{22}H_{25}NO_6$	399.43
Cyclazocine	$C_{18}H_{25}NO$	271.39
Cyclobarbital	$C_{12}H_{16}N_2O_3$	236.26
Hexodorm, Phanodorn, Sonaform		
Cyclobenzaprine	$C_{20}H_{21}N$	275.38
Cyclopenthiazide	$C_{13}H_{18}ClN_3O_4S_2$	379.89
Navidrex, Salimid		
Cyclophosphamide	$C_7H_{15}Cl_2N_2O_2P$	261.10
Cytoxan, Endoxan, Sendoxan		
Cyclothiazide	$C_{14}H_{16}ClN_3O_4S_2$	389.91
Anhydron®		
Cyproheptadine	$C_{21}H_{21}N$	287.39
Periactinol		
Dantrolene	$C_{14}H_{10}N_4O_5$	314.26
Dantrium®		
Dapsone	$C_{12}H_{12}N_2O_2S$	248.30
Avlosulfon®, Disulone, Udolac		
Daunomycin	$C_{27}H_{29}NO_{10}$	527.51
(Daunorubicin)		
Cerubidin®		
Desipramine	$C_{18}H_{22}N_2$	266.37
Dextromethorphan	$C_{18}H_{25}NO$	271.41
Agrippol, Contratuss, Sedatuss		
Diazepam	$C_{16}H_{13}ClN_2O$	284.76
Valium®, Canazipam, E-Pam, Novodipam, Neo-Calme, Serenack		
Dibenzepin	$C_{18}H_{21}N_3O$	295.37
Ansiopax, Deprex, Neodalit, Noveril		
Dibucaine	$C_{20}H_{29}N_3O_2$	343.42
(Cinchocaine)		
Nupercaine®		
Dichlorophenamide	$C_6H_6Cl_2N_2O_4S_2$	305.16
(Dichlorphenamide)		
Antidrasi, Barastonin, Daranide		
Diclofenac	$C_{14}H_{11}Cl_2NO_2$	296.13
Voltaren®, Voltarol		

Appendix (Continued)

MOLECULAR WEIGHT OF DRUGS IN TABLES

Drug Name	Molecular formula	Molecular weight
Dienestrol	$C_{18}H_{18}O_2$	266.32
Cycladiene, Dienol, Dinovex		
Diethylpropion	$C_{13}H_{19}NO$	205.30
(Amfepramone)		
Dietec, Nobesine-25, Regibon, Tenuate®		
Diethylstilbestrol	$C_{18}H_{20}O_2$	268.34
Stiblium		
Digitoxin	$C_{41}H_{64}O_{13}$	764.92
Purodigin		
Digoxin	$C_{41}H_{64}O_{14}$	780.92
Lanoxin®, Rougoxin		
Dihydralazine	$C_8H_{10}N_6$	190.21
Nepresol		
Diminazene	$C_{14}H_{15}N_7$	281.30
Diphenhydramine	$C_{17}H_{21}NO$	255.35
Benadryl®		
Diphenylhydantoin	$C_{15}H_2N_2O_2$	252.26
(Phenytoin)		
Dilantin®, Dantoin, Novodiphenyl		
Dipyridamole	$C_{20}H_{40}N_8O_4$	504.62
Persantine®		
Disopyramide	$C_{21}H_{29}N_3O$	339.47
Norpace®, Rythmodan		
L-Dopa	$C_9H_{11}NO_4$	197.19
(Levodopa)		
Dopar®, Lardopa, Levopa		
Dothiepin	$C_{19}H_{21}NS$	295.45
Prothiaden		
Doxapram	$C_{24}H_{30}N_2O_2$	378.50
Dorpam®, Stimulexin		
Doxepin	$C_{19}H_{21}NO$	279.37
Sinequan®, Aponal, Curatin		
Dyphylline	$C_{10}H_{14}N_4O_4$	254.25
(Diprophylline)		
Dilin, Protophylline		
Econazole	$C_{18}H_{15}Cl_3N_2O$	381.68
Pevanyl		
Emepronium Bromide	$C_{20}H_{28}BrN$	362.37
Cetiprin, Restenacht, Hexanium		
Emetine	$C_{29}H_{40}N_2O_4$	480.63
Enflurane	$C_3H_2ClF_5O$	184.50
Ethrane		
Ephedrine	$C_{10}H_{15}NO$	165.23
Epinephrine	$C_9H_{13}NO_3$	183.20
Ergotamine	$C_{33}H_{35}N_5O_5$	581.65
Erythromycin	$C_{37}H_{67}NO_{13}$	733.92
Emcinka Estolate, E-Mycin®, Ilotycin, Novorythro Estolate, Robimycin		
Ethambutol	$C_{10}H_{24}N_2O_2$	204.31
Etibi, Myambutol		
Ethanol	C_2H_6O	46.07
(Alcohol, Ethyl Alcohol)		
Ethchlorvynol	C_7H_9ClO	144.61
Placidyl®		
Ethinamate	$C_9H_{13}NO_2$	167.20
Valamin, Valmid®, Valmidate		

Appendix (Continued)

MOLECULAR WEIGHT OF DRUGS IN TABLES

Drug Name	Molecular formula	Molecular weight
Ethosuximide	$C_7H_{11}NO_2$	141.17
Zarontin®		
Ethotoin	$C_{11}H_{12}N_2O_2$	204.22
Ethylene Glycol	$C_2H_6O_2$	62.07
(1,2 Ethanediol, Glycol)		
Ethyl Ether	$C_4H_{10}O$	74.12
Ethylmorphine	$C_{19}H_{23}NO_3$	313.38
Etilefrin	$C_{10}H_{15}NO_2$	181.23
Fenfluramine	$C_{12}H_{16}F_3N$	231.27
Ponderal, Pondimin®		
Flufenamic Acid	$C_{14}H_{10}F_3NO_2$	281.24
Achless, Felunamin, Paraflu®		
Flunitrazepam	$C_{16}H_{12}FN_3O_3$	313.30
RO5-4200, Rohpynol		
5-Fluorocytosine	$C_4H_4FN_3O$	129.09
(Flucytosine)		
Ancobon®, Ancotil		
5-Fluorouracil	$C_4H_3FN_2O_2$	130.08
Fluril, Efudex, Fluoroplex		
Fluphenazine	$C_{22}H_{26}F_3N_3OS$	437.52
Elinol, Valamina, Vespazine		
Flurazepam	$C_{21}H_{23}ClFN_3O$	387.89
Dalmane®		
Flurbiprofen	$C_{15}H_{13}FO_2$	244.27
Froben		
Folic Acid	$C_{19}H_{19}N_7O_6$	441.40
(Vitamin Bc, Vitamin M)		
Folvite, Novofolacid		
Ftorafur	$C_8H_9FN_2O_3$	200.16
Futraful		
Furazolidone	$C_8H_7N_3O_5$	225.16
Furoxone®		
Furosemide	$C_{12}H_{11}ClN_2O_5S$	330.77
Furoside, Lasix®, Uritol		
Gentamicin	Mixture	
(Gentamycin)		
Glibenclamide	$C_{23}H_{28}ClN_3O_5S$	494.00
(Glyburide, Glybenzcylclamide)		
Daonil, Dibeta, Euglucon		
Glutethimide	$C_{13}H_{15}NO_2$	217.26
Doriden®, Elrodorm, Doriden-Sed®		
Glycerol	$C_3H_8O_3$	92.09
(Trihydroxypropane)		
Gramicidin	Mixture	
Griseofulvin	$C_{17}H_{17}ClO_6$	352.77
Fulvicin-U/F, Grisovin		
Guanethidine	$C_{10}H_{22}N_4$	198.31
Ismelin®		
Haloperidol	$C_{21}H_{23}ClFNO_2$	375.88
Haldol®		
Halothane	$C_2HBrClF_3$	197.39
Fluothane®		
Heptabarbital	$C_{13}H_{18}N_2O_3$	250.29
Heptadorm, Medomin		
Heroin	$C_{21}H_{23}NO_5$	369.40
(Diacetylmorphine, Diamorphine)		

Appendix (Continued)

MOLECULAR WEIGHT OF DRUGS IN TABLES

Drug Name	Molecular formula	Molecular weight
Hexadiphane	$C_{21}H_{27}N$	293.43
(Prozapine)		
Norbiline		
Hexobarbital	$C_{12}H_{16}N_2O_3$	236.26
Evipal		
4-Hexylresorcinol	$C_{12}H_{18}O_2$	194.26
Caprokol		
Homatropine	$C_{16}H_{21}NO_3$	275.33
Mesopin, Novatropine		
Hydralazine	$C_8H_8N_4$	160.18
Apresoline®, Depressan		
Hydrochlorothiazide	$C_7H_8CLN_3O_4S_2$	297.72
Esidrix®, Hydrid, Hydro-Aquil, Hydro-Diuril®, Novohydrazide, Urozide		
Hydrocodone	$C_{18}H_{21}NO_3$	299.36
Bekadid, Dicodid®		
Hydroflumethiazide	$C_8H_8F_3N_3O_4S_2$	331.29
Diucardin		
p-Hydroxybenzoic Acid	$C_7H_6O_3$	138.13
Ibuprofen	$C_{13}H_{18}O_2$	206.27
Motrin®		
Imipramine	$C_{19}H_{24}N_2$	280.40
Tofranil®, Impramil, Impril, Novopramine, Praminil		
Indomethacin	$C_{19}H_{16}ClNO_4$	357.81
Indocid, Infrocin, Indocin®		
Indoprofen	$C_{17}H_{15}NO_3$	281.32
Flosin, Flosint, Isindone		
Iodochlorhydroxyquin	C_9H_5ClINO	305.52
Cliquinol, Entero-Vioform, Vioform		
Iproclozide	$C_{11}H_{15}ClN_2O_2$	242.72
Sursum		
Iproniazid	$C_9H_{13}N_3O$	179.22
Euphozid, Marsilid		
Isofluorphate	$C_6H_{14}FO_3P$	184.15
(Difluorophate, Dyflos Floropryl)		
Isoniazid	$C_6H_7N_3O$	137.15
(INH, Isonicotinic Acid Hydrazide)		
Isotamine, Rimifon		
Isosorbide Dinitrate	$C_6H_8N_2O_8$	236.14
(Sorbide Nitrate)		
Isordil®, Coronex		
Kanamycin	Mixture	
Ketamine	$C_{13}H_{16}ClNO$	237.74
Ketalar®		
Ketoprofen	$C_{16}H_{14}O_3$	259.29
Alrheumat, Orudis, Profenid		
Lasalocid	$C_{34}H_{54}O_8$	590.80
Avatec		
Lidocaine	$C_{14}H_{22}N_2O$	234.33
(Lignocaine)		
Xylocaine®		
Lorazepam	$C_{15}H_{10}Cl_2N_2O_2$	321.16
Ativan®, Emotival, Lorax		
Loxapine	$C_{18}H_{18}ClN_3O$	327.81
Lysergide	$C_{20}H_{25}N_3O$	323.42
(LSD, LSD-25)		

Appendix (Continued)

MOLECULAR WEIGHT OF DRUGS IN TABLES

Drug Name	Molecular formula	Molecular weight
Delysid		
Mafenide	$C_7H_{10}N_2O_2S$	186.25
Sulfamylon, Marfanil, Napaltan		
Maprotiline	$C_{20}H_{23}N$	277.41
Ludiomil®		
Mebendazole	$C_{16}H_{13}N_3O_3$	295.30
Vermox®		
Meclizine	$C_{25}H_{27}ClN_2$	390.96
Bonamine®		
Medazepam	$C_{16}H_{15}ClN_2$	270.76
Nobrium®		
Mefenamic Acid	$C_{15}H_{15}NO_2$	241.28
Ponstan®		
Mefruside	$C_{13}H_{19}ClN_2O_5S_2$	382.90
Baycaron®		
Melphalan	$C_{13}H_{18}Cl_2N_2O_2$	305.20
Alkeran®		
Meperidine	$C_{15}H_{21}NO_2$	247.35
(Pethidine)		
Demerol®		
Mephenytoin	$C_{12}H_{14}N_2O_2$	218.25
Mesantoin®		
Mephobarbital	$C_{13}H_{14}N_2O_3$	246.26
(Methylphenobarbital)		
Mebaral®, Isonal®		
Mepirizole	$C_{11}H_{14}N_4O_2$	234.26
Mebron®		
Meprobamate	$C_9H_{18}N_2O_4$	218.25
Equanil®, Miltown®, Mep-E, Neo-Tan, Novomepro		
6-Mercaptopurine	$C_5H_4N_4S$	152.19
Purinethol		
Metformin	$C_4H_{11}N_5$	129.17
Glucophage		
Methadone	$C_{21}H_{27}NO$	309.45
Amidone®		
Methamphetamine	$C_{10}H_{15}N$	149.24
Norodin®		
Methapyrilene	$C_{14}H_{19}N_3S$	261.38
M-P, Histadyl®, Lullamin		
Methaqualone	$C_{16}H_{14}N_2O$	250.29
Mequelon, Quaalude®, Sedalone, Tualone-300		
Methimazole	$C_4H_6N_2S$	114.17
Tapazole®		
Methotrexate	$C_{20}H_{22}N_8O_5$	454.46
A-Methopterin		
Methotrimeprazine	$C_{19}H_{24}N_2OS$	328.46
(Levomepromazine)		
Methoxyflurane	$C_3H_4Cl_2F_2O$	164.97
Metofane, Penthrane, Pentrane		
Methoxyphenamine	$C_{11}H_{17}NO$	179.25
Orthoxine		
8-Methoxypsoralen	$C_{12}H_8O_4$	216.18
(Methoxalen)		
Meloxine, Oxsoralen, Methoxa-Dome		
Methyldopa	$C_{10}H_{13}NO_4$	211.21
Aldomet®, Dopamet, Novomedopa		

Appendix (Continued)

MOLECULAR WEIGHT OF DRUGS IN TABLES

Drug Name	Molecular formula	Molecular weight
Methylglucamine	$C_7H_{17}NO_5$	195.21
(Meglumine)		
Methylphenidate	$C_{14}H_{19}NO_2$	233.30
Ritalin®, Methidate		
Methylsalicylate	$C_8H_8O_3$	152.14
Methyprylon	$C_{10}H_{17}NO_2$	183.26
Noludar®		
Metoclopramide	$C_{14}H_{22}ClN_3O_2$	299.81
Maxeran®		
Metolazone	$C_{16}H_{16}ClN_3O_3S$	365.84
Zaroxolyn		
Metronidazole	$C_6H_9N_3O_3$	171.16
Flagyl, Neo-Tric, Novonidazol, Trikacide		
Mianserin	$C_{18}H_{20}N_2$	264.37
Tolvin		
Miconazole	$C_{18}H_{14}Cl_4N_2O$	416.12
Micatin, Monistat®		
Morazone	$C_{23}H_{27}N_3O_2$	377.47
Tarugan		
Morphine	$C_{17}H_{19}NO_3$	285.33
Nafiverine	$C_{34}H_{38}N_2O_4$	538.66
Naftidan		
Nalidixic Acid	$C_{12}H_{12}N_2O_3$	232.23
NegGram®, Nogram, Wintomylon		
Naloxone	$C_{19}H_{21}NO_4$	327.37
Narcon		
Naproxen	$C_{14}H_{14}O_3$	230.26
Naprosyn®		
Nefopam	$C_{17}H_{19}NO$	253.35
Acupan		
Neomycin	Mixture	
Mycifradin, Myciguent, Neocin		
Neostigmine Bromide	$C_{12}H_{19}BrN_2O_2$	303.20
Prostigmin® bromide		
Neostigmine Methyl Sulfate	$C_{13}H_{22}N_2O_6S$	334.39
Prostigmine® methylsulfate		
Nicotine	$C_{10}H_{14}N_2$	162.23
Nicotinic Acid	$C_6H_5NO_2$	123.11
(Niacin)		
Nicamin, Niconacid, Nico-Span		
Nifedipine	$C_{17}H_{18}N_2O_6$	346.34
Adalat		
Nikethamide	$C_{10}H_{14}N_2O$	178.23
Coramine®, Kardonyl		
Nimorazole	$C_9H_{14}N_4O_3$	226.23
Acterol		
Niridazole	$C_6H_6N_4O_3S$	214.22
Ambilhar		
Nitrazepam	$C_{15}H_{11}N_3O_3$	281.26
Megadon, Mogadon®		
Nitrofurantoin	$C_8H_6N_4O_5$	238.16
Furanex, Furatine, Macrodantin®, Nephronex, Novofuran		
Nitroglycerin	$C_3H_5N_3O_9$	227.09
(Glyceryl Trinitrate)		
Nitrong, Nitrostabilin, Nitrostat		

Appendix (Continued)

MOLECULAR WEIGHT OF DRUGS IN TABLES

Drug Name	Molecular formula	Molecular weight
Nortriptyline	$C_{19}H_{21}N$	263.37
Aventyl®		
Novobiocin	$C_{31}H_{36}N_2O_{11}$	612.65
Albamycin, Cardelmycin		
Nylidrin	$C_{19}H_{25}NO_2$	299.40
Arlidin®, Pervadil		
Orphenadrine	$C_{18}H_{23}NO$	269.37
Disipal, Norflex®, Mephenamine		
Oxazepam	$C_{15}H_{11}ClN_2O_2$	286.74
Serax®		
Oxprenolol	$C_{15}H_{23}NO_3$	265.34
Oxycodone	$C_{18}H_{21}NO_4$	315.36
Oxyphenbutazone	$C_{19}H_{20}N_2O_3$	324.37
Tandearil®		
Oxyphenonium Bromide	$C_{21}H_{34}BrNO_3$	428.41
Antrenyl		
Pancuronium Bromide	$C_{35}H_{60}Br_2N_2O_4$	732.70
Pavulon®		
Papaverine	$C_{20}H_{21}NO_4$	339.38
Paraldehyde	$C_6H_{12}O_3$	132.16
(Paracetaldehyde)		
Pargyline	$C_{11}H_{13}N$	159.22
Eutonyl®		
Pemoline	$C_9H_8N_2O_2$	176.17
Deltamine, Kethamed, Pioxol, Stimul, Volital		
Penicillamine	$C_5H_{11}NO_2S$	149.21
Cuprimine		
Penicillin G	$C_{16}H_{18}N_2O_4S$	334.8
(Benzyl Penicillin)		
Penicillin V	$C_{16}H_{18}N_2O_5S$	350.38
Pentazocine	$C_{19}H_{27}NO$	285.44
Talwin®, Fortral		
Pentobarbital	$C_{11}H_{18}N_2O_3$	226.26
Nembutal®		
Pentylenetetrazole	$C_6H_{10}N_4$	138.17
(Pentetrazol)		
Pentrazol		
Perazine	$C_{20}H_{25}N_3S$	339.49
(Pemazine)		
Psytomin, Taxilan		
Perphenazine	$C_{21}H_{26}ClN_3OS$	403.97
Phenazine, Trilafon®		
Phanquone	$C_{12}H_6N_2O_3$	210.18
Entobex		
Phenacetin	$C_{10}H_{13}NO_2$	179.21
(Acetophenetidin)		
Phencyclidine	$C_{17}H_{25}N$	243.38
(PCP)		
Sernyl		
Phendimetrazine	$C_{12}H_{17}NO$	191.26
Dietrol, Plegine®		
Pheneformin	$C_{10}H_{15}N_5$	205.27
DBI®, Dibotin, Insoral		
Phenobarbital	$C_{12}H_{12}N_2O_3$	232.23
Eskabarb, Gardenal, Luminal, Mediphen, Nova-Pheno		

Appendix (Continued)

MOLECULAR WEIGHT OF DRUGS IN TABLES

Drug Name	Molecular formula	Molecular weight
Phenol (Carbolic Acid)	C_6H_6O	94.11
Phenprocoumon Marcumar	$C_{18}H_{16}O_3$	280.31
Phentermine Ionamin®	$C_{10}H_{15}N$	149.23
Phenylbutazone Butazolidin®, Phenbutazone, Novophenyl	$C_{19}H_{20}N_2O_2$	308.37
Phenylephrine Neo-Synephrine®	$C_9H_{13}NO_2$	167.17
Phenylpropanolamine Coldecon	$C_9H_{13}NO$	151.17
Pilocarpine Adsorbocarpine, Nova-Carpine, Pentacarpine	$C_{11}H_{16}N_2O_2$	208.25
Piperazine Antepar®, Entacyl, Piperzinal	$C_4H_{10}N_2$	86.14
Piracetam Euvifor, Normabrain	$C_6H_{10}N_2O_2$	142.15
Polythiazide Renese®	$C_{11}H_{13}Cl F_3N_3O_4S_3$	439.90
Practolol Dalzic,Eradin®	$C_{14}H_{22}N_2O_3$	264.32
Prazepam Demetrin, Verstran®	$C_{19}H_{17}ClN_2O$	342.83
Prazosin Hypovase, Minipress	$C_{19}H_{21}N_5O_4$	382.42
Primidone (primaclone) Mysoline®, Sertan	$C_{12}H_{14}N_2O_2$	218.25
Probencid Benuryl	$C_{13}H_{19}NO_4S$	285.36
Procainamide Ponestyl®, Novocamid	$C_{13}H_{21}N_3O$	235.29
Procaine Novocain®	$C_{13}H_{20}N_2O_2$	236.30
Procarbazine Natulan	$C_{12}H_{19}N_3O$	221.30
Propantheline Bromide Pro-Banthine®, Propanthel, Novopropanthil	$C_{23}H_{30}BrNO_3$	448.42
Propionylpromazine Combelene	$C_{20}H_{24}N_2OS$	340.55
Propoxyphene (Dextropropoxyphene) Darvon®, Depronal SA, Pro-65	$C_{22}H_{29}NO_2$	339.48
Propranolol Inderal®	$C_{16}H_{21}NO_2$	259.34
Propylthiouracil Propyl-Thyracil, Thyreostat II	$C_7H_{10}N_2OS$	170.23
Propyphenazone	$C_{14}H_{18}N_2O$	230.30
Protriptyline Triptil	$C_{19}H_{21}N$	263.37
Proxyphylline Brontyl, Purophyllin, Theon	$C_{10}H_{14}N_4O_3$	238.24
Pyrazinamide Tebrazid	$C_5H_5N_3O$	123.11

Appendix (Continued)

MOLECULAR WEIGHT OF DRUGS IN TABLES

Drug Name	Molecular formula	Molecular weight
Pyridinol Carbamate	$C_{11}H_{15}N_3O_4$	253.25
Anginin		
Pryidoxine	$C_8H_{11}NO_3$	169.18
(Vitamin B_6)		
Hexa-Betalin, Hexa-Vibex, Winvite-6		
Pyrimethamine	$C_{12}H_{13}ClN_4$	248.71
Daraprim®		
Pyrithioxin	$C_{16}H_{20}N_2O_4S_2$	368.48
Enerbol		
Pyrithyldione	$C_9H_{13}NO_2$	167.20
Persedon, Benedorm		
Quinidine	$C_{20}H_{24}N_2O_2$	324.41
Klinidin, Klinidrin, Quiniduran, Quinate		
Reserpine	$C_{33}H_{40}N_2O_9$	608.70
Neo-Serp, Resercrine, Serpasil®		
Ribavirin	$C_8H_{12}N_4O_5$	244.21
Virazole®		
Riboflavine	$C_{17}H_{20}N_4O_6$	376.36
(Vitamin B_2)		
Flamotide, Hibon, Viras		
Rifampicin	$C_{43}H_{58}N_4O_{12}$	822.96
(Rifampin)		
Rifadin®, Rimactane®		
Saccharin	$C_7H_5NO_3S$	183.18
Salbutamol	$C_{13}H_{21}NO_3$	239.31
(Albuterol)		
Ventolin®		
Salicylamide	$C_7H_7NO_2$	137.13
Dolomide Salymid		
Salicylanilide	$C_{13}H_{11}NO_2$	213.24
Salinidol		
Salicylic Acid	$C_7H_6O_3$	138.23
Secobarbital	$C_{12}H_{18}N_2O_3$	228.27
(Quinalbarbital)		
Seconal®		
Sorbitol	$C_6H_{14}O_6$	182.17
(D-Glucitol)		
Spectinomycin	$C_{14}H_{24}N_2O_7$	332.35
Trobicin®		
Spiramycin	Mixture	
Rovamycine		
Streptozocin	$C_8H_{15}N_3O_7$	265.22
Strychnine	$C_{21}H_{22}N_2O_2$	334.40
Succinylcholine Bromide	$C_{14}H_{30}Br_2N_2O_4$	450.23
(Succinyldicholine Bromide, Suxamethonium Bromide)		
Sulbenicillin	$C_{16}H_{18}N_2O_7S_2$	414.45
Kedacillin Lilacillin		
Sulfabenzamide	$C_{13}H_{12}N_2O_3S$	276.31
Sulfachloropyridazine	$C_{10}H_9ClN_4O_2S$	284.74
Sulfadiazine	$C_{10}H_{10}N_4O_2S$	250.28
Sulfadimethoxine	$C_{12}H_{14}N_4O_4S$	310.33
Sulfamerazine	$C_{11}H_{12}N_4O_2S$	264.30
Sulfamethazine	$C_{12}H_{14}N_4O_2S$	278.32
Sulfamethoxypyridazine	$C_{11}H_{12}N_4O_3S$	280.32

Appendix (Continued)

MOLECULAR WEIGHT OF DRUGS IN TABLES

Drug Name	Molecular formula	Molecular weight
Sulfapyridine	$C_{11}H_{11}N_3O_2S$	249.29
Sulfaquinoxaline	$C_{14}H_{12}N_4O_2S$	300.33
Sulfathiazole	$C_9H_9N_3O_2S_2$	255.32
Sulfinpyrazone	$C_{23}H_{20}N_2O_3S$	404.48
Anturan®		
Sulfisoxazole	$C_{11}H_{13}N_3O_3S$	267.30
Sulthiame	$C_{10}H_{14}N_2O_4S_2$	290.37
(Sultiame)		
Ospolot		
Synephrine	$C_9H_{13}NO_2$	167.20
(Oxedrine)		
Analeptin, Vasocordrin		
Tamoxifen	$C_{26}H_{29}NO$	371.53
Novadex		
Terbutaline	$C_{12}H_{19}NO_3$	225.29
Bricanyl®, Brethine®		
Tetracycline	$C_{22}H_{24}N_2O_8$	444.43
Achromycin®, Cefracycline, Novotetra, Sunycin, Tetracyn		
Tetramisole	$C_{11}H_{12}N_2S$	204.31
Anthelvet, Citarin, Nilverm, Orovermol, Spartakon		
Theophylline	$C_7H_8N_4O_2$	180.17
Aminophyllin, Elixophyllin®, Theolixir, Oxtriphylline, Acet-Am, Theolamine		
Thiamine	$C_{12}H_{17}ClN_4OS$	300.78
(Vitamin B_1)		
Becrinol, Betazin, Vervon		
Thiazinamium Methylsulfate	$C_{19}H_{26}N_2O_4S_2$	410.55
Mulltergan, Padisal, Valan		
Thimerosal	$C_9H_9HgNaO_2S$	404.84
(Thiomersal)		
Merthiolate		
Thiopental	$C_{11}H_{18}N_2O_2S$	242.33
(Thiopentane)		
Pentothal		
Thioridazine	$C_{21}H_{26}N_2S_2$	370.56
Mellaril®, Novoridazine, Thioril		
Thiothixene	$C_{23}H_{29}N_3O_2S_2$	443.62
Navane®		
Thyroxine	$C_{15}H_{11}I_4NO_4$	776.93
(Dextrothyroxine, Levothyroxine)		
Tilidine	$C_{17}H_{23}NO_2$	273.38
Perdolat		
Timolol	$C_{13}H_{24}N_4O_3S$	316.42
Blocadren®		
Tinidazole	$C_8H_{13}N_3O_4S$	247.26
Fasigyne, Simplotan, Tricolam		
Tolazamide	$C_{14}H_{21}N_3O_3S$	311.41
Diabewas, Norglycin, Tolinase		
Tolbutamide	$C_{12}H_{18}N_2O_3S$	270.34
Mobenol, Novobutamide, Orinase®, Oramide		
Tolmetin	$C_{15}H_{15}NO_3$	257.30
Tolectin®		
Tranylcypromine	$C_9H_{11}N$	133.19
Parnate®, Tylciprine		
TRH	$C_{16}H_{22}N_6O_4$	362.40
(Thyrotropin-Releasing Hormone)		
Thypinone		

Appendix (Continued)

MOLECULAR WEIGHT OF DRUGS IN TABLES

Drug Name	Molecular formula	Molecular weight
Triameterene Dyrenium®	$C_{12}H_{11}N_7$	253.26
Tribromsalan Diaphene, Temasept IV, Tualsal 100	$C_{13}H_8Br_3NO_2$	449.96
Triclocarban	$C_{13}H_9Cl_3N_2O$	315.59
Trihexylphenidyl (Benzhexol) Aparkane, Artane®, Novohexidyl	$C_{20}H_{31}NO$	301.46
Trimethadione (Troxidone) Trimedone	$C_6H_9NO_3$	143.14
Trimethoprim Syraprim	$C_{14}H_{18}N_4O_3$	290.32
Trimipramine Surmontil®, Stangyl	$C_{20}H_{26}N_2$	294.42
Triprolidine Actidil®	$C_{19}H_{22}N_2$	278.38
Tris(hydroxymethyl) aminomethane (Tromethamine) Tris, Tham	$C_4H_{11}NO_3$	121.14
Valproic Acid Depakene®, Epilim	$C_8H_{16}O_2$	144.21
Verapamil Cordilox	$C_{27}H_{38}N_2O_4$	454.59
Vidarabine Vira-A	$C_{10}H_{13}N_5O_4$	267.26
Vinbarbital Delvinal	$C_{11}H_{16}N_2O_3$	224.26
Vincamine	$C_{21}H_{26}N_2O_3$	354.43
Vitamin A (Retinol)	$C_{20}H_{30}O$	286.44
Vitamin B_{12} (Cyanocobalamin, Cobalamin)	$C_{63}H_{88}CoN_{14}O_{14}P$	1355.42
Vitamin C (L-Ascorbic Acid)	$C_6H_8O_6$	176.12
Vitamin D;2 (Calciferol)	$C_{28}H_{44}O$	396.63
Vitamin D_3 (7-Dehydrocholesterol)	$C_{27}H_{44}O$	384.62
Vitamin E (α-Tocopherol)	$C_{29}H_{50}O_2$	430.69
Vitamin K_1 (Menadione, Menaphthone, Phytomendione)	$C_{31}H_{46}O_2$	450.68
Warfarin Coumadin®	$C_{19}H_{16}O_4$	308.32

INDEX

A

B

C

D

E

F

G

H

I

J

K

L

M

N

O

P

Q

R

S

T

U

V

W

X

Y

Z